New Advances in Oil, Gas and Geothermal Reservoirs

New Advances in Oil, Gas and Geothermal Reservoirs

Editor

Daoyi Zhu

MDPI • Basel • Beijing • Wuhan • Barcelona • Belgrade • Manchester • Tokyo • Cluj • Tianjin

Editor
Daoyi Zhu
China University of
Petroleum-Beijing at Karamay
China

Editorial Office
MDPI
St. Alban-Anlage 66
4052 Basel, Switzerland

This is a reprint of articles from the Special Issue published online in the open access journal *Energies* (ISSN 1996-1073) (available at: https://www.mdpi.com/journal/energies/special_issues/Oil_Gas_Geothermal_Reservoirs).

For citation purposes, cite each article independently as indicated on the article page online and as indicated below:

LastName, A.A.; LastName, B.B.; LastName, C.C. Article Title. *Journal Name* **Year**, *Volume Number*, Page Range.

ISBN 978-3-0365-6402-9 (Hbk)
ISBN 978-3-0365-6403-6 (PDF)

Contents

About the Editor

Daoyi Zhu

Daoyi Zhu graduated from Yangtze University with a Bachelor's of Science degree in Applied Chemistry in 2011, and received a Master's degree in Chemical Engineering and a Ph.D. in Oil and Gas Field Development Engineering from China University of Petroleum (Beijing) in 2014 and 2018, respectively. In 2018, he joined the Petroleum School of China University of Petroleum-Beijing at Karamay, and was a visiting scholar at the Department of Petroleum Engineering, Missouri University of Science and Technology from September 2016 to March 2018. He is currently an Associate Professor of Petroleum Engineering (since 2021). In addition, he is an associate editor of *Petroleum Science*, and serves as guest editor and executive editorial board member of other international journals. Dr. Zhu focuses on oilfield chemistry, reservoir conformance improvement, CCUS and geo-energy development. He has published more than 30 peer-reviewed journal articles, two textbooks, *Oilfield Chemistry* and *Literature Search and Technical Writing*, and numerous translation books.

 energies

MDPI

Editorial

New Advances in Oil, Gas, and Geothermal Reservoirs

Daoyi Zhu

Petroleum School, China University of Petroleum-Beijing at Karamay, Karamay 834000, China; zhudaoyi@cupk.edu.cn

Abstract: The most significant geo-energy sources in the world today continue to be oil, gas, and geothermal reservoirs. To increase oil and gas reserves and production, new theories are constantly being developed in the laboratory and new technologies are being applied in the oilfield. This Special Issue compiles recent research focusing on cutting-edge ideas and technology in oil, gas, and geothermal reservoirs, covering the fields of well drilling, cementing, hydraulic fracturing, improved oil recovery, conformance control, and geothermal energy development.

Keywords: drilling and completion; oil well cement; fracturing; tight gas; shale gas; enhanced thermal recovery; enhanced geothermal system

1. Introduction

The most significant geo-energy sources worldwide in recent years continue to be oil, natural gas, and geothermal resources. As conventional fossil energy resources, oil and gas still play an irreplaceable role in industrial production and human life. In addition, with the continued recovery of the global economy and increasing energy consumption, the development of oil and natural gas is still very important.

The development of crude oil and natural gas mainly comprises the stages of drilling, completion, well cementing, fracturing and acidification, and improving oil recovery (IOR) [1]. With rapid developments in experimental technology and numerical simulation technology, drilling speed increases and drilling becomes safer, and the quality of well cementing gradually increases. Fracturing and acidification are two main measures to increase the production for low- and ultra-low-permeability reservoirs [2]. The stimulated reservoir volume (SRV) has been continuously improved, and the proppant placement scheme can be continuously optimized. An eternal theme of the development of oil fields, IOR technologies are gradually applied to unconventional reservoirs, such as heavy- and tight-oil reservoirs [3]. In addition, new types of plugging materials have been developed to overcome the low sweep efficiency of the displacement agent caused by reservoir heterogeneity [4–6].

Geothermal resources, a kind of renewable thermal energy inside the Earth with huge reserves, are considered a feasible alternative to achieving global dual carbon goals. The key issue to resolve is the use of science and technology to improve the recovery efficiency of geothermal resources.

This collection, which accompanies the Special Issue of *Energies*, places an emphasis on fundamental innovations and compiles 11 current publications on original applications of new ideas and methodologies in oil, gas, and geothermal reservoirs.

Citation: Zhu, D. New Advances in Oil, Gas, and Geothermal Reservoirs. *Energies* **2023**, *16*, 477. https://doi.org/10.3390/en16010477

Received: 13 December 2022
Accepted: 25 December 2022
Published: 1 January 2023

2. Review of Research Presented in This Special Issue

The papers published in this Special Issue describe recent advancements in oil, gas, and geothermal reservoirs. These studies are divided into five categories.

The first category is the efficient and high-speed drilling and completion technology of complex hydrocarbon formations. In response to the situation where the cuttings removal mechanism of the current pulsed jet is unclear, Zhao et al. [7] established a pressure-flow-rate fluctuation model and discussed the impact of the displacement, drilling fluid viscosity, well depth, and flowing area of the pulsed jet tool on the instant flow characteristics at the bottom hole. The findings indicated that flow velocity fluctuations can improve the mechanical status of the cuttings, and help cuttings to tumble off the bottom hole.

The quality of well cementing also determines the quality of the later development process. As the main material for well cementing, oil well cement has been widely studied in recent years. Qi et al. [8] used low-field pulse nuclear magnetic resonance (LF-NMR) technology to examine the impact of the addition of concrete retarders on T_2 distribution, cement thickening characteristics, and paste strength. According to studies, retarders can slow the rate at which cement particles hydrate by weakening the van der Waals force and the electrostatic absorption force between water and cement grains. The early strength of cement slurry can be enhanced by appropriately increasing the cation content of polymeric ion retarders.

The second category is the low-cost and highly efficient development of low/ultra-low permeability hydrocarbon reservoirs. Due to the low matrix permeability, these reservoirs are generally required to employ a fracturing method to increase the flow volume of oil and gas [9]. The purpose of the fracturing process is to pump the proppant into the fractures, causing the oil and gas in the low penetration substrate to easily flow out [10]. Wu et al. [11] examined proppant transportation and placement in narrow curving channels. The dune height and covered area of the proppant in the curving fractures are smaller than in linear fractures. Additionally, curving sections hinder the distribution of the proppant, making the location of the dunes closer to the inlet.

For the hydraulic fracture technology in glutenite reservoirs, the geometric heterogeneity of the embedded gravel affects its serious stress and strength heterogeneities. Tang et al. [12] used the discrete element method (DEM) method to simulate the macro mechanical behavior of gravel samples. Results showed that gravel embedded near the wellbore can cause stress and strength heterogeneities, which further increase the local initial point and form a complex fracture network nearby.

In addition, imbibition or gas flooding methods are usually used after fracturing. Deng et al. [13] used a microfluidic platform based on a visual circular capillary tube to monitor the entire imbibition process. Studies found that when analyzing the mathematical models of the imbibition process, the impact of the collected liquid ejecting from the capillary tube on the imbibition must be considered. This effect is more obvious in the imbibition process of lower viscosity (e.g., water) liquid displacing higher viscosity liquid (e.g., oil). The imbibition behavior in the capillary was described using additional force, and a revised Poiseuille mathematical model was developed. The model can make excellent predictions on the water imbibition process.

The third category is the effective development of tight gas and shale gas reservoirs. Research on the existence and flowing characteristics of natural gas (mainly methane) in tight or shale reservoirs is essential for the reserve assessment and production prediction of tight and shale gas. To explore the effects of methane's adsorption dynamics on gas shale rocks, the methane adsorption dynamic law on gas shale powder was described by Zhang et al. [14]. Most methane molecules are adsorbed during the early stage, and low temperature is conducive to the adsorption of methane on shale powder.

The inorganic nanopores in the matrix of shale form irreducible water on their inner surfaces due to strong hydrophilicity. By establishing a corrected shale gas apparent permeability model, Zhao et al. [15] analyzed the impact of the existence of irreducible water on natural gas flow behavior. Studies showed that the rate of bulk phase transportation replacing the surface diffusion of natural gas slowed down because of the existence of irreducible water. However, with the increase in formation pressure, the impact of irreducible water on the apparent permeability of gas decreases. Under low-pressure conditions, the irreducible water in the small pores can accelerate the flow of natural gas.

The fourth category is the further improving oil recovery (IOR) technology of the mature oil fields. Although they gradually entered the post-development period, and the efficiency of water injection gradually decreased, there is still a large amount of remaining oil in the formation [16]. Therefore, novel IOR technology for old oilfields has always been a hotspot for oil companies and researchers. Cyclic supercritical multithermal fluid stimulation (CSMTFS) is a new type of technology for the development of heavy oil reservoirs. Tian et al. [17] conducted a simulation experiment on its heating efficiency, production, and heat loss through a three-dimensional (3D) experimental system. Results indicated that CSMTF had less heat loss than conventional thermal fluids, and the enthalpy values are significantly increased compared to multithermal fluids; therefore, it can improve the heat-carrying ability of the multithermal fluid. Tang et al. [18] discussed the characteristics of supercritical multithermal fluids and their potential in enhanced thermal recovery. The experimental results show that the oil-to-water ratio of the reactant has a more significant impact on the specific enthalpy and displacement efficiency compared with the initial temperature and pressure. With a low gas-to-water ratio, the supercritical multithermal fluid had a higher crude oil displacement efficiency and oil recovery in the beginning and enlarged the supercritical area; therefore, later channels can be formed in the oil layer.

Aiming at the problems caused by low steam sweep efficiency in heavy oil reservoirs, Cheng et al. [19] used flour ash as a plugging material to evaluate its controlling steam channeling in the two-dimensional (2D) displacement experiment. It has good anti-flush ability, stable plugging performance, and excellent improved heavy oil recovery effects.

The fifth category is the efficient development of geothermal resources. Due to the continuous increase in geothermal resources in recent years, and the growing difficulty of conventional fossil energy development, the efficient utilization of geothermal resources is increasingly valuable. To improve the recharge efficiency of geothermal resources, Yu et al. [20] proposed an unblocking and permeability enhancement method using a rotary water jet for low recharge efficiency wells in sandstone geothermal reservoirs. This technology can solve the problem of low efficiency in the reinjection processes of cooled thermal waters back into geothermal reservoirs.

3. Conclusions

Many academics from a variety of fields, from the natural sciences to engineering, have been conducting research into drilling and completion, the development of conventional and unconventional oil and gas reservoirs, IOR technology of mature oil fields, and the development of geothermal resources. New theories and technologies are proposed in this Special Issue. The new experimental methods, numerical simulation technology, and pilot cases in this Special Issue can help readers and researchers to better understand and be inspired by the cutting-edge technologies in the field of oil and gas and geothermal.

Acknowledgments: With the gracious assistance and support of the editorial staff, the guest editors would like to thank MDPI for the opportunity to serve as the guest editors of this Special Issue of *Energies*. The guest editors are also appreciative of the authors' inspirational contributions and the enormous work of the anonymous reviewers.

Conflicts of Interest: The authors declare no conflict of interest.

References

1. Zhu, D. *Oilfield Chemistry*; Petroleum Industry Press: Beijing, China, 2020.
2. Zhang, H.; Zhu, D.; Gong, Y.; Qin, J.; Liu, X.; Pi, Y.; Zhao, Q.; Luo, R.; Wang, W.; Zhi, K. Degradable preformed particle gel as temporary plugging agent for low-temperature unconventional petroleum reservoirs: Effect of molecular weight of the cross-linking agent. *Petrol. Sci.* **2022**, *19*, 3182–3193. [CrossRef]
3. Zhu, D.; Zhao, Y.; Zhang, H.; Zhao, Q.; Shi, C.; Qin, J.; Su, Z.; Wang, G.; Liu, Y.; Hou, J. Combined imbibition system with black nanosheet and low-salinity water for improving oil recovery in tight sandstone reservoirs. *Petrol. Sci.* **2022**. Available online: https://doi.org/10.1016/j.petsci.2022.12.004 (accessed on 30 December 2022).
4. Zhu, D.; Deng, Z.; Chen, S. A review of nuclear magnetic resonance (NMR) technology applied in the characterization of polymer gels for petroleum reservoir conformance control. *Petrol. Sci.* **2021**, *18*, 1760–1775. [CrossRef]

5. Zhu, D.; Wang, Z.; Liu, Y.; Zhang, H.; Qin, J.; Zhao, Q.; Wang, G.; Shi, C.; Su, Z. Carbon Nanofiber-Enhanced HPAM/PEI Gels for Conformance Control in Petroleum Reservoirs with High Temperatures. *Energy Fuel* **2022**, *36*, 12606–12616. [CrossRef]
6. Zhu, D.; Bai, B.; Hou, J. Polymer gel systems for water management in high-temperature petroleum reservoirs: A chemical review. *Energy Fuel* **2017**, *31*, 13063–13087. [CrossRef]
7. Zhao, H.; Shi, H.; Huang, Z.; Chen, Z.; Gu, Z.; Gao, F. Mechanism of Cuttings Removing at the Bottom Hole by Pulsed Jet. *Energies* **2022**, *15*, 3329. [CrossRef]
8. Qi, Z.; Chen, Y.; Yang, H.; Gao, H.; Hu, C.; You, Q. An NMR Investigation of the Influence of Cation Content in Polymer Ion Retarder on Hydration of Oil Well Cement. *Energies* **2022**, *15*, 8881. [CrossRef]
9. Zhao, S.; Zhu, D.; Bai, B. Experimental study of degradable preformed particle gel (DPPG) as temporary plugging agent for carbonate reservoir matrix acidizing to improve oil recovery. *J. Petrol. Sci. Eng.* **2021**, *205*, 108760. [CrossRef]
10. Wu, G.; Chen, S.; Meng, X.; Wang, L.; Sun, X.; Wang, M.; Sun, H.; Zhang, H.; Qin, J.; Zhu, D. Development of antifreeze fracturing fluid systems for tight petroleum reservoir stimulations. *Energy Fuel* **2021**, *35*, 12119–12131. [CrossRef]
11. Wu, Z.; Wu, C.; Zhou, L. Experimental Study of Proppant Placement Characteristics in Curving Fractures. *Energies* **2022**, *15*, 7169. [CrossRef]
12. Tang, J.; Liu, B.; Zhang, G. Investigation on the Propagation Mechanisms of a Hydraulic Fracture in Glutenite Reservoirs Using DEM. *Energies* **2022**, *15*, 7709. [CrossRef]
13. Deng, W.; Liang, T.; Yuan, S.; Zhou, F.; Li, J. Abnormal Phenomena and Mathematical Model of Fluid Imbibition in a Capillary Tube. *Energies* **2022**, *15*, 4994. [CrossRef]
14. Zhang, Y.; Hu, A.; Xiong, P.; Zhang, H.; Liu, Z. Experimental Study of Temperature Effect on Methane Adsorption Dynamic and Isotherm. *Energies* **2022**, *15*, 5047. [CrossRef]
15. Zhao, S.; Liu, H.; Jiang, E.; Zhao, N.; Guo, C.; Bai, B. Study on Apparent Permeability Model for Gas Transport in Shale Inorganic Nanopores. *Energies* **2022**, *15*, 6301. [CrossRef]
16. Deng, Z.; Liu, M.; Qin, J.; Sun, H.; Zhang, H.; Zhi, K.; Zhu, D. Mechanism study of water control and oil recovery improvement by polymer gels based on nuclear magnetic resonance. *J. Petrol. Sci. Eng.* **2022**, *209*, 109881. [CrossRef]
17. Tian, J.; Yan, W.; Qi, Z.; Huang, S.; Yuan, Y.; Dong, M. Cyclic Supercritical Multi-Thermal Fluid Stimulation Process: A Novel Improved-Oil-Recovery Technique for Offshore Heavy Oil Reservoir. *Energies* **2022**, *15*, 9189. [CrossRef]
18. Tang, X.; Hua, Z.; Zhang, J.; Fu, Q.; Tian, J. A Study on Generation and Feasibility of Supercritical Multi-Thermal Fluid. *Energies* **2022**, *15*, 8027. [CrossRef]
19. Cheng, K.; Liu, Y.; Qi, Z.; Tian, J.; Luo, T.; Hu, S.; Li, J. Laboratory Evaluation of the Plugging Performance of an Inorganic Profile Control Agent for Thermal Oil Recovery. *Energies* **2022**, *15*, 5452. [CrossRef]
20. Yu, C.; Tian, T.; Hui, C.; Huang, H.; Zhang, Y. Study on Unblocking and Permeability Enhancement Technology with Rotary Water Jet for Low Recharge Efficiency Wells in Sandstone Geothermal Reservoirs. *Energies* **2022**, *24*, 9407. [CrossRef]

Article

Mechanism of Cuttings Removing at the Bottom Hole by Pulsed Jet

Heqian Zhao, Huaizhong Shi *, Zhongwei Huang, Zhenliang Chen, Ziang Gu and Fei Gao

State Key Laboratory of Petroleum Resources and Prospecting, China University of Petroleum (Beijing), Beijing 102249, China; 2017312014@student.cup.edu.cn (H.Z.); huangzw@cup.edu.cn (Z.H.); 2020310128@student.cup.edu.cn (Z.C.); 2019215147@student.cup.edu.cn (Z.G.); 2020215159@student.cup.edu.cn (F.G.)

* Correspondence: shz@cup.edu.cn; Tel.: +86-01089733512

Abstract: Vibration drilling technology induced by hydraulic pulse can assist the bit in breaking rock at deep formation. Simultaneously, the pulsed jet generated by the hydraulic pulse promotes removal of the cuttings from the bottom hole. Nowadays, the cuttings removal mechanism of the pulsed jet is not clear, which causes cuttings to accumulate at the bottom hole and increases the risk of repeated cutting. In this paper, a pressure-flow rate fluctuation model is established to analyze the fluctuation characteristics of the pulsed jet at the bottom hole. Based on the model, the effects of displacement, well depth, drilling fluid viscosity, and flow area of the pulsed jet tool on the feature of instantaneous flow at the bottom hole are discussed. The results show that the pulsed jet causes flow rate and pressure to fluctuate at the bottom hole. When the displacement changes from 20 L/s to 40 L/s in a 2000 m well, the pulsed jet generates a flow rate fluctuation of 4–9 L/s and pressure fluctuation of 0.1–0.5 MPa at the bottom hole. With the flow area of the tool increasing from 2 cm^2 to 4 cm^2, the amplitude of flow rate fluctuation decreases by 72.5%, and the amplitude of pressure fluctuation decreases by more than 60%. Combined with the fluctuation feature of the flow field and the water jet attenuation law at the bottom hole, the force acting on the cuttings under the pulsed jet is derived. It is found that flow rate fluctuation improves the mechanical state of cuttings and is beneficial for cuttings tumbled off the bottom hole. This research provides theoretical guidance for pulsed jet cuttings cleaning at the bottom hole.

Keywords: drilling; unconventional oil and gas; cuttings cleaning; pulsed jet

Citation: Zhao, H.; Shi, H.; Huang, Z.; Chen, Z.; Gu, Z.; Gao, F. Mechanism of Cuttings Removing at the Bottom Hole by Pulsed Jet. *Energies* **2022**, *15*, 3329. https://doi.org/10.3390/en15093329

Academic Editors: Daoyi Zhu and Rouhi Farajzadeh

Received: 24 March 2022
Accepted: 29 April 2022
Published: 3 May 2022

1. Introduction

With the consumption of conventional fossil energy, unconventional oil and gas have become more and more important. In addition to hydraulic fracturing and horizontal wells, increasing the rate of penetration (ROP) is essential for the efficient development of unconventional oil and gas resources [1–3]. As an efficient rock-breaking method, the water jet has gradually been applied in mining and oil drilling [4]. With in-depth study of the water jet, it is found that pulsed jet shows a better rock-breaking effect than continuous jet.

There are two methods to modulate a continuous jet into a pulsed jet. One is to change the instantaneous flow mechanically, by adjusting the flow area to generate pressure fluctuations and pulsed jet. The other is to design the channel structure based on principles of hydroacoustics to modulate continuous jets into a pulsed jet [5]. In the 1980s, the pulsed jet nozzle was developed and applied to the drill bit [6]. The pulsed jet nozzle can form a small flow area and then form a high jet impact force, which is helpful for wellbore cleaning [6,7]. Compared to the cone-straight nozzle, the erosion effect of the pulse nozzle is significantly improved. Through investigation of the pulsed jet nozzle, it is found that the pulsed jet nozzle can increase instantaneous dynamic pressure at the bottom hole, which is conducive to cuttings migration at the bottom hole. Due to the small size of the nozzle, it generates a pulsed jet with relatively weak energy at the bottom hole. Therefore, many scholars began to develop pulsed jet tools. Tempress Technology Company of the United States developed the HydroPulse negative pressure pulse tool [8]. Through field experiments, it was found that the ROP could be increased by 33% to 200%. The principle of the tool is to control the valve switch by the drilling fluid, which causes the fluid to flow discontinuously. The discontinuous flow forms a pulsed jet at the bottom hole. The Waltech Company designed a negative pressure pulse tool based on change of fluid flow direction through the Coanda principle [9,10]. It generates a pulsed jet at the bottom hole to enhance the ROP. The hydraulic pulsed cavitating jet-assisted drilling tool was invented by the China University of Petroleum [11]. The tool has been popularized and widely applied in oil and gas fields in China and abroad. It has achieved an obvious effect on improving the ROP [12,13].

Predecessors have carried out some research on rock-breaking properties of the pulsed jet. Experimental results indicate that pulsed jet has a stronger eroding capacity than continuous jet [14,15]. In order to visualize the pulsed jet flow field, shadowgraph technology, combined with particle image velocimetry (PIV) method, is used to capture the flow field feature of the pulsed jet [16]. In order to obtain the fluctuation feature of the pulsed jet at the bottom hole, a numerical model is established, which is based on unstable flow in the wellbore [17,18]. In addition, the fluctuation of the pressure at the bottom hole is important for an oil well. On the one hand, it provides many pieces of information related to the formation in the well test [19–21]. On the other hand, it improves the magnitude of the cutting at the bottom in drilling.

Many scholars have carried out abundant research on the rock-breaking and ROP enhancing effect of the pulsed jet. However, research on cuttings cleaning of the pulsed jet at the bottom hole is still relatively lacking. In this paper, a pressure-flow rate fluctuation model at the bottom hole is established. The method of characteristics is adopted to solve the model. Based on the feature of pressure-flow rate fluctuation and water jet attenuation law at the bottom hole, the force acted on the cuttings at the bottom hole is analyzed. The effects of different displacements, well depths, drilling fluid viscosities, and flow areas of pulse jet tools on instantaneous flow and pressure at the bottom hole in pulse jet drilling are analyzed. In addition, the influence of different bottom hole pressure differences and cutting sizes on the force of cuttings is analyzed. The paper attempts to describe the cuttings clearing mechanism at the bottom hole of the pulsed jet with a quantitative method.

2. Modeling of Pressure-Flow Rate Fluctuation at the Bottom Hole

2.1. Modeling

Figure 1 shows the conventional drilling process. In a drilling process, the drill fluid flows through the kelly, drill pipe, and collar to the bit in sequence. After the fluid flows through the nozzles on the bit, the jet is formed at the bottom hole. The jet cleans the cuttings at the bottom hole and assists the bit in breaking the rock. Then, the fluid flows outside of the well head through the annulus, carrying the cuttings. For pulsed jet drilling, a pulsed jet generator is often installed above the drill bit to modulate the continuous flow into a pulsed jet at the bottom hole. Due to complex drilling conditions, there are some reasonable assumptions in the model:

(1) It is assumed that the wellbore is vertical.
(2) It is assumed that the size of the drill pipe is identical. The flow characteristics at the joint are ignored.
(3) It is assumed that the drilling string contains only the drill pipe, not the collars and other bottom hole assembly.
(4) The difference between the inner and outer diameter of the pulse jet generator and the drill pipe is ignored.
(5) The flow in the wellbore is one-dimensional. The radial flow in the drill pipe and annulus is ignored.

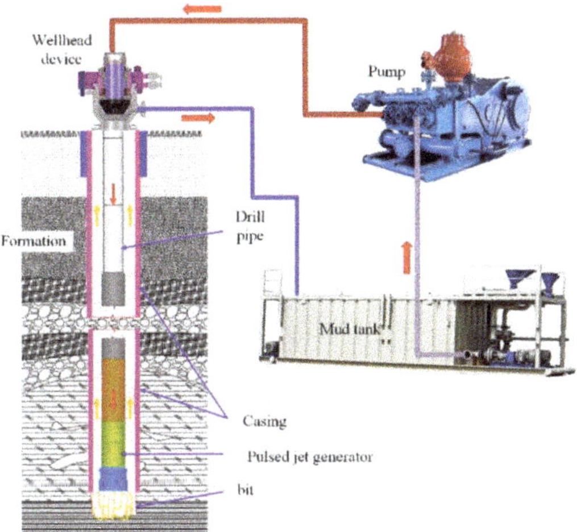

Figure 1. Schematic of drilling.

Figure 2 shows a schematic of the fluid flow in the wellbore during the drilling process. Based on the control volume 1—1—2—2 shown in Figure 2, the mass conservation equation and momentum equation in the drill pipe are established, respectively, as shown in Equations (1) and (2).

$$\frac{\partial}{\partial x}(\rho v A_p) + \frac{\partial}{\partial t}(\rho A_p) = 0 \tag{1}$$

$$\frac{\partial v}{\partial t} + v\frac{\partial v}{\partial x} + \frac{1}{\rho}\frac{\partial p}{\partial x} + 2\frac{\tau_p}{\rho r_{pi}} - g = 0 \tag{2}$$

where $A_p = \pi {r_{pi}}^2$ is the flow area of the drill pipe, v is the fluid velocity, ρ is the density of the fluid, τ_p is the shear stress on the inner wall of the drill pipe, r_{pi} is the radius of the drill pipe.

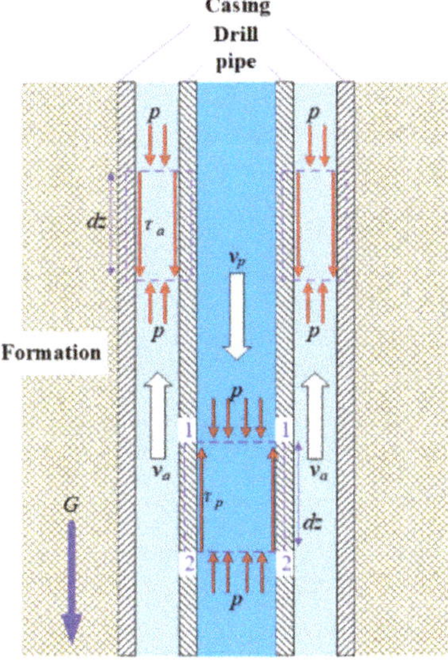

Figure 2. Fluid flow in the wellbore.

Expanding Equation (1) and introducing the full derivative equation, Equation (3) can be derived:

$$\frac{1}{A_p}\frac{dA_p}{dt} + \frac{1}{\rho}\frac{d\rho}{dt} + \frac{\partial v}{\partial x} \tag{3}$$

The first term and the second term on the left side of Equation (3) are respectively related to the variation rate of the pipe section and the density of the fluid, which are caused by the pressure variation of the elastic wave. The variation rates are relative to the compressibility coefficient of the fluid and the expansion coefficient of the flow channel:

$$\alpha = \frac{\frac{dV}{V}}{dp}, \beta = \frac{\frac{dA_p}{A_p}}{dp} \tag{4}$$

where α is the compressibility coefficient of the fluid and β is the expansion coefficient of the flow channel.

According to the total derivative of p, substitute Equation (4) into (3):

$$(\alpha + \beta)\frac{dp}{dt} + \frac{\partial v}{\partial x} = 0 \Leftrightarrow v\frac{\partial v}{\partial x} + \frac{\partial p}{\partial t} + \frac{1}{\alpha + \beta}\frac{\partial v}{\partial x} = 0 \tag{5}$$

The fluid compressibility coefficient and the flow channel expansion coefficient affect the propagation velocity of the pressure wave C in the pipe as follows:

$$\rho C^2 = \frac{1}{\alpha + \beta} \tag{6}$$

Combining Equations (5) and (6), the fluid continuity equation in the pipeline under the condition of pressure fluctuation is obtained:

$$v\frac{\partial v}{\partial x} + \frac{\partial p}{\partial t} + \rho C^2\frac{\partial v}{\partial x} = 0 \tag{7}$$

In Equation (2), $2\frac{\tau_p}{\rho r_{pi}}$ is the pressure loss caused by friction between the fluid and the inner wall of the drill pipe, which is related to the flow velocity and the pipe diameter, which is equal to $\frac{f_a v|v|}{2D}$, where f_a is the Darcy-Weisbach friction factor, and D is the inner diameter of the drill pipe.

For unsteady flow, the friction factor contains the effect of steady flow and fluid fluctuations [22]:

$$f_a = f_q + \frac{kD}{v|v|}\left[\frac{\partial v}{\partial t} + C\mathrm{sign}(v)\left|\frac{\partial v}{\partial x}\right|\right] \tag{8}$$

where k is the Brunone friction coefficient, and sign is the sign function.

$$k = \frac{\sqrt{C^*}}{2} \tag{9}$$

where C^* is the Vardy shear attenuation coefficient [23]:

$$\begin{cases} C^* = 0.00476 & \text{Laminar flow} \\ C^* = \frac{7.41}{\mathrm{Re}^{\log(14.3/\mathrm{Re}^{0.05})}} & \text{Turbulence flow} \end{cases} \tag{10}$$

Through derivation, the governing equations of fluid flow, considering pressure fluctuation, are obtained:

$$\begin{cases} \frac{\partial p}{\partial t} + v\frac{\partial p}{\partial x} + \rho c^2\frac{\partial v}{\partial x} = 0 \\ \frac{\partial v}{\partial t} + v\frac{\partial v}{\partial x} + \frac{1}{\rho}\frac{\partial p}{\partial x} - g + \frac{fv|v|}{2D} + k\left[\frac{\partial v}{\partial t} + \mathrm{sign}(v)c\left|\frac{\partial v}{\partial x}\right|\right] = 0 \end{cases} \tag{11}$$

2.2. Model Solving

The governing equations in this paper are composed of the mass conservation equation and the momentum equation of the fluid. It is difficult to obtain the exact solution of the equation by the analytical method. Therefore, the numerical solution method is adopted. Through comparison, it is found that the mass conservation equation and the momentum equation can be transformed into a typical hyperbolic differential equation, and the partial differential equation can be transformed into an ordinary differential equation, using the method of characteristics (MOC). Compared with the direct finite difference equation, the MOC method has high solution accuracy and is easy to program [17,18].

According to the MOC method, the mass conservation equation and the momentum equation are linearly combined by the coefficient λ:

$$\begin{aligned} \frac{\partial v}{\partial t} + v\frac{\partial v}{\partial x} + \frac{1}{\rho}\frac{\partial p}{\partial x} - g + \frac{fv|v|}{2D} + k\left[\frac{\partial v}{\partial t} + sign(v)c\left|\frac{\partial v}{\partial x}\right|\right] + \\ \lambda\left(\frac{\partial p}{\partial t} + v\frac{\partial p}{\partial x} + \rho c^2\frac{\partial v}{\partial x}\right) = 0 \end{aligned} \tag{12}$$

The total derivatives of velocity and pressure to time are introduced as:

$$\begin{cases} \frac{dp}{dt} = \frac{\partial p}{\partial t} + \frac{\partial p}{\partial x}\frac{dx}{dt} \\ \frac{dv}{dt} = \frac{\partial v}{\partial t} + \frac{\partial v}{\partial x}\frac{dx}{dt} \end{cases} \tag{13}$$

It is found that the terms on the left side of Equation (12) can be combined into a form of total derivatives of velocity and pressure, as follows:

$$\begin{cases} (1+k)\frac{\partial v}{\partial t} + \left[v + k\mathrm{sign}\left(v\frac{\partial v}{\partial z}\right)c + \lambda\rho c^2\right]\frac{\partial v}{\partial z} = (1+k)\left[\frac{\partial v}{\partial t} + \frac{B}{1+k}\frac{\partial v}{\partial z}\right] \\ \frac{1}{\rho}\frac{\partial p}{\partial z} + \lambda\frac{\partial p}{\partial t} + \lambda v\frac{\partial p}{\partial z} = \lambda\left[\frac{\partial p}{\partial t} + \frac{D}{\lambda}\frac{\partial p}{\partial z}\right] \end{cases} \tag{14}$$

where $B = v + k\mathrm{sign}\left(v\frac{\partial v}{\partial z}\right)c + \lambda\rho c^2$, $D = \frac{1}{\rho} + \lambda v$.

When the coefficient λ satisfies Equation (15), the linear combination of the mass conservation equation and the momentum equation is transformed into an ordinary differential equation, shown as Equation (16).

$$\rho c^2\lambda^2 + [E - (1+k)v]\lambda - (1+k)\frac{1}{\rho} = 0 \tag{15}$$

where $E = v + k\mathrm{sign}\left(v\frac{\partial v}{\partial x}\right)c$.

$$(1+k)\frac{dv}{dt} + \lambda\frac{dp}{dt} - g + \frac{fv|v|}{2D} = 0 \tag{16}$$

The value of λ can be calculated from Equation (15):

$$\lambda^{\pm} = \frac{(1+k)v - E \pm \sqrt{[E-(1+k)v]^2 + 4\rho c^2(1+k)\frac{1}{\rho}}}{2\rho c^2} \tag{17}$$

In order to solve the equation, a grid of space x and time t is established, as shown in Figure 3. The red dotted line LN is the forward characteristic line, corresponding to the previous characteristic equation in Equation (18); the blue dotted line RN is the backward characteristic line, corresponding to the latter characteristic equation in Equation (18).

$$\begin{cases} \frac{dx}{dt}^{+} = \frac{1}{\rho\lambda^{+}} + v \Rightarrow \frac{dt}{dx}^{+} = \frac{\rho\lambda^{+}}{1+\rho\lambda^{+}v} \\ \frac{dx}{dt}^{-} = \frac{1}{\rho\lambda^{-}} + v \Rightarrow \frac{dt}{dx}^{-} = \frac{\rho\lambda^{-}}{1+\rho\lambda^{-}v} \end{cases} \tag{18}$$

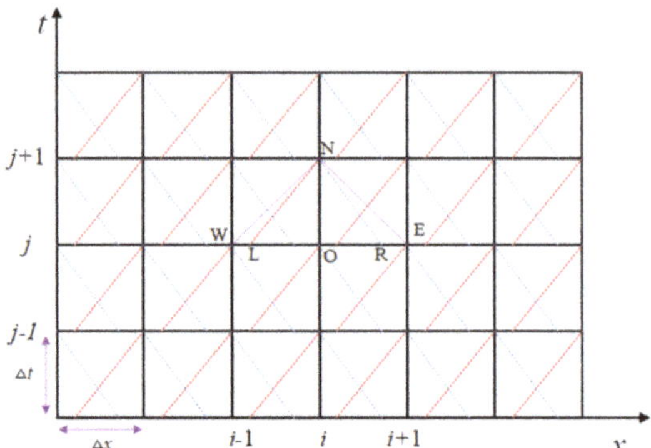

Figure 3. Schematic of MOC method.

As shown in Figure 3, since point N satisfies the two characteristic line equations of LN and RN, the equations, including the pressure and velocity of points L, N, and R, are established based on Equation (16). The first equation in Equation (19) corresponds to the characteristic line LN, and the second equation corresponds to the characteristic line RN.

$$\begin{cases} (1+k_N)(v_N - v_L) + \lambda_N{}^{+}(p_N - p_L) + \left(\frac{f v_N |v_N|}{2D} - g\right)\triangle t = 0 \\ (1+k_N)(v_N - v_R) + \lambda_N{}^{-}(p_N - p_R) + \left(\frac{f v_N |v_N|}{2D} - g\right)\triangle t = 0 \end{cases} \tag{19}$$

The pressure and velocity relationship between the point N and points L and R can be derived from Equation (19), as shown in Equation (20). In addition, in each time step, the pressure and velocity at points L and R can be calculated from the values of points W and E using linear interpolation. Therefore, an explicit relationship of the governing equations at every point in Figure 3 is obtained using the MOC method.

$$\begin{cases} p_N = \frac{(1+k_N)(v_R - v_L) - \lambda_N{}^{+} p_L + \lambda_N{}^{-} p_R}{(\lambda_N{}^{-} - \lambda_N{}^{+})} \\ v_N = v_L - \lambda_N{}^{+} \frac{(1+k_N)(v_R - v_L) - \lambda_N{}^{+} p_L + \lambda_N{}^{-} p_R}{(\lambda_N{}^{-} - \lambda_N{}^{+})(1+k_N)} + \frac{\lambda_N{}^{+} p_L}{(1+k_N)} \\ -\left(\frac{f v_N |v_N|}{2D} - g\right)\frac{\triangle t}{(1+k_N)} \end{cases} \tag{20}$$

2.3. *Initial and Boundary Conditions*

Figure 4 shows the boundary conditions and the distribution of spatial nodes during the model solution process. Based on the assumption mentioned above, the boundary conditions include the drill string inlet (mud pump outlet), the pulse jet tool and drill bit, and the annular outlet (mud tank).

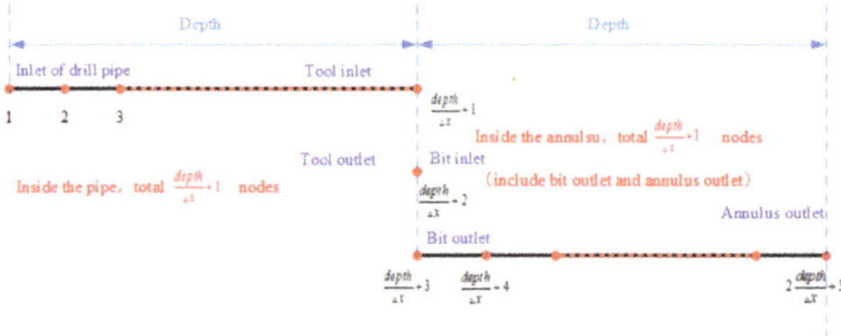

Figure 4. Schematic diagram of boundary and space nodes.

The boundary condition at the drill string inlet is the outlet of the mud pump. The mud pump is set to a constant displacement Q_0. The pressure at the inlet of the drill string changes with pressure fluctuations in the wellbore. Based on the MOC method, the pressure at the inlet can be calculated as follows:

$$\begin{cases} Q_1^{j+1} = Q_0 \Leftrightarrow v_1^{j+1} = \frac{Q_0}{A_a} \\ p_R = p_1^j - \frac{1+\rho\lambda_1^{j+1-} v_1^{j+1}}{\rho\lambda_1^{j+1-}} \frac{\Delta t}{\Delta x}\left(p_2^j - p_1^j\right) \\ v_R = v_1^j - \frac{1+\rho\lambda_1^{j+1-} v_1^{j+1}}{\rho\lambda_1^{j+1-}} \frac{\Delta t}{\Delta x}\left(v_2^j - v_1^j\right) \\ p_1^{j+1} = p_R - \frac{\left(1+k_1^{j+1}\right)\left(v_1^{j+1} - v_R\right) + \left(\frac{f v_1^{j+1}\left|v_1^{j+1}\right|}{2D} - g\right)\Delta t}{\lambda_1^{j+1-}} \end{cases} \quad (21)$$

The boundary condition at the tool and the bit are relative to the pressure drop and instantaneous flow rate. The pressure drop of the tool is equal to the pressure difference between the points $\frac{depth}{\Delta x} + 1$ and $\frac{depth}{\Delta x} + 3$, which can be obtained by the MOC method, as shown in Figure 4. Based on the relationship between the pressure drop and flow rate at the tool and bit, the fluid velocity of the pulsed jet tool can be calculated:

$$v_{tool} = \frac{-B + \sqrt{B^2 - 4AC}}{2A} \quad (22)$$

where

$$\begin{aligned} A &= \frac{f_a\left(\frac{S}{A_a}\right)^2 \Delta t}{2D_a \lambda_a{}^-} - \frac{f_p\left(\frac{S}{A_p}\right)^2 \Delta t}{2D_p \lambda_p{}^+} - \frac{1}{2}\rho\left(1 - \frac{S}{A_p}\right)\left(\frac{3}{2} - \frac{S}{A_p}\right) - \frac{1}{2}\rho\left(\frac{S}{CA_n}\right)^2 \\ B &= \frac{(1+k_a)\frac{S}{A_a}}{\lambda_a{}^-} - \frac{(1+k_p)\frac{S}{A_p}}{\lambda_p{}^+} \\ C &= p_L - p_R + \frac{(1+k_p)v_L}{\lambda_p{}^+} + \frac{g\Delta t}{\lambda_p{}^+} - \frac{(1+k_a)v_R}{\lambda_a{}^-} + \frac{g\Delta t}{\lambda_a{}^-} \end{aligned} \quad (23)$$

where the subscripts a and p correspond to the annuls and pipe, respectively. S is the flow area of the tool. The instantaneous pressure and flow rate of the tool and bit can be calculated from the velocity at the tool.

The pressure at the annulus outlet is 0 Pa, which is atmospheric pressure. The velocity is obtained by the MOC method, as shown in Equation (24).

$$\begin{cases} p^{j+1}_{2\frac{depth}{\Delta z}+3} = 0 \\ v_L = v^j_{2\frac{depth}{\Delta z}+2} - \frac{1+\rho\lambda^{j+1\,+}_{2\frac{depth}{\Delta z}+3} v^{j+1}_{2\frac{depth}{\Delta z}+3}}{\rho\lambda^{j+1\,+}_{2\frac{depth}{\Delta z}+3}} \frac{\Delta t}{\Delta z}\left(v^j_{2\frac{depth}{\Delta z}+2} - v^j_{2\frac{depth}{\Delta z}+3}\right) \\ p_L = p^j_{2\frac{depth}{\Delta z}+2} - \frac{1+\rho\lambda^{j+1\,+}_{2\frac{depth}{\Delta z}+3} v^{j+1}_{2\frac{depth}{\Delta z}+3}}{\rho\lambda^{j+1\,+}_{2\frac{depth}{\Delta z}+3}} \frac{\Delta t}{\Delta z}\left(p^j_{2\frac{depth}{\Delta z}+2} - p^j_{2\frac{depth}{\Delta z}+3}\right) \\ v^{j+1}_{2\frac{depth}{\Delta z}+3} = v_L - \frac{\lambda^{j+1\,+}_{2\frac{depth}{\Delta z}+3}\left(p^{j+1}_{2\frac{depth}{\Delta z}+3} - p_L\right) + \left(\frac{f v^{j+1}_{2\frac{depth}{\Delta z}+3}\left|v^{j+1}_{2\frac{depth}{\Delta z}+3}\right|}{2D_a} + g\right)\Delta t}{1 + k^{j+1}_{2\frac{depth}{\Delta z}+3}} \end{cases} \quad (24)$$

At the initial time, the flow rate in the drill pipe and annulus is set as Q_0. Moreover, the pulsed jet tool is in a state of minimum flow area.

Figure 5 is the flowchart of solving the model. First, it is necessary to input the basic parameters and the initial and boundary conditions. Then, the step size for space and time is selected. The pressure and flow rate at points L and R in Figure 3 are calculated using the interpolation method. Based on the MOC method, the pressure and flow rate of each point at the next time step are calculated. Due to the update of the values of each point, it is necessary to evaluate the convergence of the pressure and flow rate. Finally, the results are obtained when the iteration of all the points is finished.

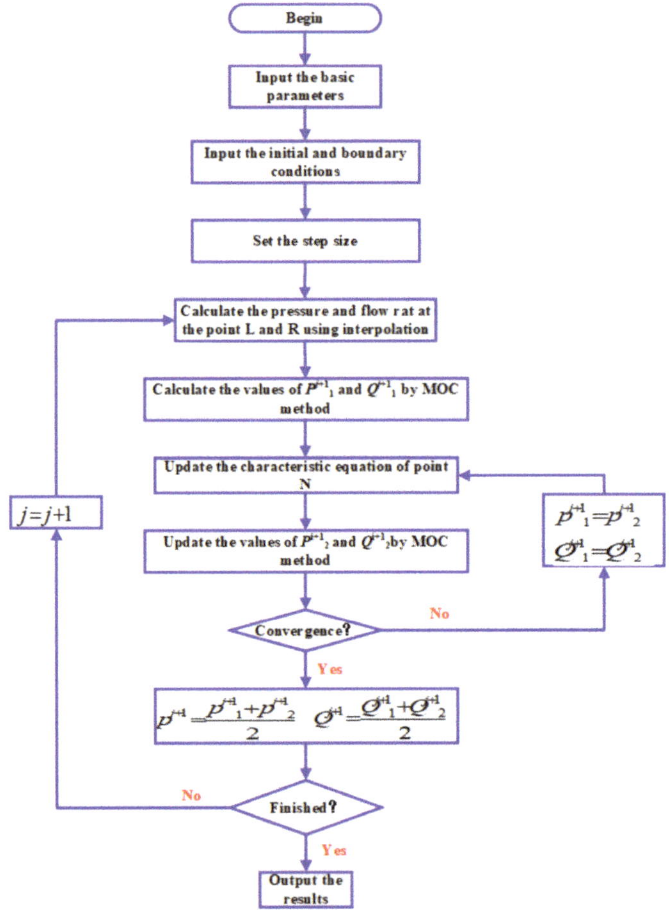

Figure 5. Model solving flowchart.

3. Cuttings Cleaning Model with Pulsed Jet

During the drilling process, the PDC bit teeth break the rock and generate fractures at the bottom hole. In addition, the pulsed jet vibration drilling technic promotes cracks by the vibration of the bit at the bottom hole [24,25]. Under the water jet, the cuttings escape from the bottom hole into the annulus and are carried out of the wellbore. For deep wells and ultra-deep wells, huge drilling fluid column pressure creates a chip hold-down effect, which prevents cuttings from breaking away from the bottom hole, resulting in repeated cutting of the drill bit, aggravation of the bit wear and reduction of the ROP. The pulsed jet can generate pressure and flow rate fluctuations at the bottom hole, and can reduce instantaneous bottom hole pressure and alleviate the chip hold-down effect [12,26].

In order to quantify the cuttings cleaning ability of the pulsed jet, it is necessary to establish a mechanical model of the cuttings under the pulsed jet at the bottom hole. Based on the analysis mentioned above, it assumes that the cuttings are spherical. The pressure gradient and collision between the cuttings are ignored. Figure 6 shows the force for a single particle at the bottom hole. As shown, the cuttings are to be held down in the fractures at the bottom hole. According to the assumptions, the drill fluid column pressure p_c combined with the pore pressure generates the chip hold-down effect on the cuttings. The forces acting on the cuttings include gravity G, buoyancy F_b (Equation (25)), pressure force F_c and F_p (Equation (26)), and the drag force of the water jet F_D (Equation (27)). As shown in Figure 6, the nozzle forms a certain angle with the bottom hole. The drag force can be divided into horizontal and vertical directions, corresponding to F_{Dh} and F_{Dv}, respectively. According to Newton's third law, cuttings particles are also subjected to support force F_N and friction force F_f from the bottom hole.

$$\begin{cases} G = \frac{1}{6}\pi {d_p}^3 \rho_p g \\ F_b = \frac{1}{6}\pi {d_p}^3 \rho_l g \end{cases} \tag{25}$$

$$\begin{cases} F_c = p_a s_p \\ F_p = p_p s_p \end{cases} \tag{26}$$

$$F_D = \frac{1}{6}\pi {d_p}^3 \rho_p \frac{\vec{u} - \vec{u}_p}{\tau_r} \tag{27}$$

where d_p is the diameter of the cuttings, ρ_p and ρ_l are the density of cuttings and fluid, τ_r is the relaxation time, which can be calculated from Equation (28) [27]:

$$\tau_r = \frac{\rho_p {d_p}^2}{18\mu} \frac{24}{C_D \mathrm{Re}_p} \tag{28}$$

where μ is the viscosity of the fluid, C_D is the drag coefficient, Re_p is the Reynolds number of the particle.

$$\mathrm{Re}_p = \frac{\rho_l d_p \left| \vec{u} - \vec{u}_p \right|}{\mu} \tag{29}$$

where $\vec{u}$ is the fluid velocity at the bottom, which can be calculated from the water jet attenuation law, and $\vec{u}_p$ is the velocity of the cuttings, which is zero at the bottom hole.

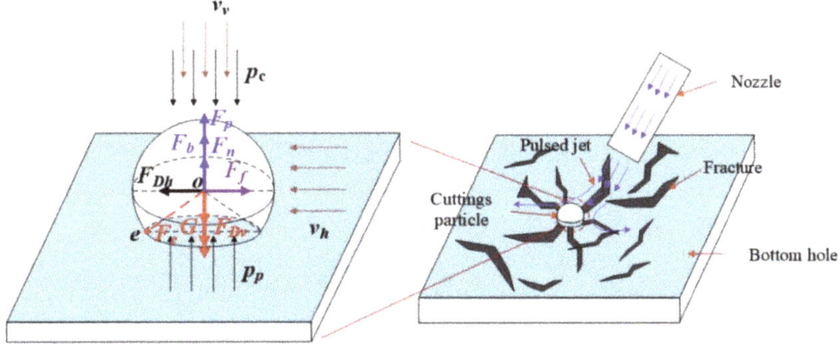

Figure 6. Schematic diagram of the force for a single particle at the bottom hole.

In this paper, the Morsi and Alexander drag model [28] is adopted to calculate the drag force of the cuttings:

$$C_D = a_1 + \frac{a_2}{\text{Re}_p} + \frac{a_3}{\text{Re}_p{}^2} \tag{30}$$

where a_1, a_2, a_3 are the coefficients corresponding to the particle Reynolds number, shown in Table 1.

Table 1. The values of coefficient for spherical drag model. Adapted with permission from Elsevier, 2022 [29].

The Range of Re_p	a_1	a_2	a_3
$\text{Re}_p < 0.1$	0	24.0	0
$0.1 < \text{Re}_p < 1.0$	3.69	22.73	0.0903
$1.0 < \text{Re}_p < 10.0$	1.222	29.1667	−3.8889
$10.0 < \text{Re}_p < 100.0$	0.6167	46.5	−116.67
$100.0 < \text{Re}_p < 1000.0$	0.3644	98.33	−2778
$1000.0 < \text{Re}_p < 5000.0$	0.357	148.62	-4.75×10^4
$5000.0 < \text{Re}_p < 10{,}000.0$	0.46	−490.546	57.87×10^4
$10{,}000.0 < \text{Re}_p < 50{,}000.0$	0.5191	−1662.5	5.4167×10^6

As shown in Figure 7, all of the forces are decomposed into the F_2 along the *oe* direction and the force F_1 perpendicular to the *oe* direction. Since F_2 and *oe* are collinear, they cannot generate the moment to point *e*. When $F_1 > 0$ and $F_2 > 0$, the cuttings particles begin to tumble off the bottom hole. When $F_2 < 0$, the cuttings break away directly from the bottom hole.

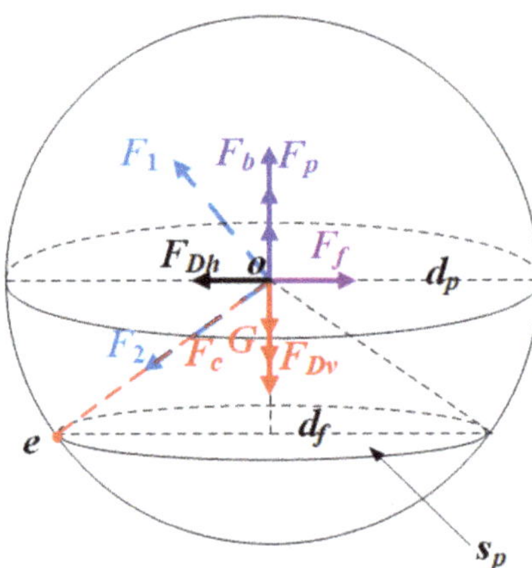

Figure 7. Schematic diagram of a single particle of cuttings.

Therefore, a mechanical model of the critical state, when cuttings particles just leave the bottom hole is analyzed. In the critical state, the support force F_N and the friction force F_f are equal to 0. Then, component force F_1 and F_2 can be obtained:

$$\begin{cases} F_1 = F_{Dh}\frac{\sqrt{d_p{}^2 - d_f{}^2}}{d_p} + (F_b + F_p - G - F_c - F_{Dv})\frac{d_f}{d_p} \\ F_2 = F_{Dh}\frac{d_f}{d_p} + (G + F_c + F_{Dv} - F_b - F_p)\frac{\sqrt{d_p{}^2 - d_f{}^2}}{d_p} \end{cases} \tag{31}$$

4. Results and Discussion

4.1. The Fluctuation Characteristics of Pressure and Flow Rate in the Wellbore

The fluctuation of pressure and flow rate in the wellbore with the 127 mm drill pipe is calculated. During the solution, the space step is set as 10 m. The time step is calculated from the Courant-Friedrichs-Lewy condition to keep the solution stable [30].

Figure 8 shows the pressure and flow rate fluctuation propagation in the wellbore, including the state at 0 s, 0.05 s, 0.1 s, 0.5 s, and 1 s. It can be found that at the initial moment, the fluid flows without fluctuation. As the pulsed jet generator starts to work, pressure fluctuations are created at the tool, which induces the flow rate fluctuation. Within the drill pipe, the pressure and flow rate fluctuations propagate along the drill pipe to the wellhead. In addition, at the rear end of the tool, the fluctuations of pressure and flow rate are transmitted through the nozzle into the annulus and propagate to the wellhead. It is found that the drilling process is stable flow without the tool. When the pulse jet tool is installed at the bottom hole, the stable flow is modulated into a pulsed jet with fluctuations.

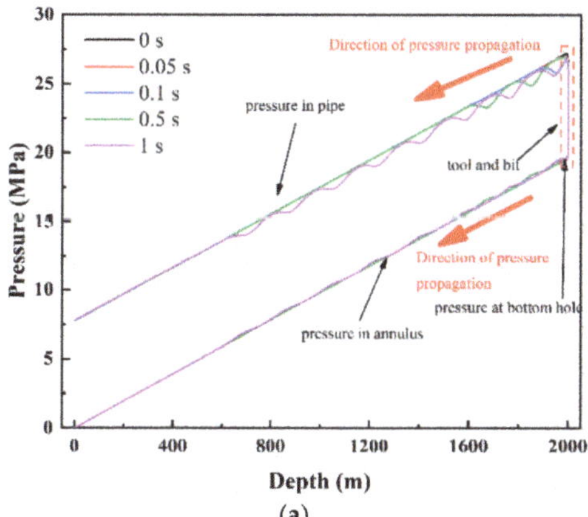

(a)

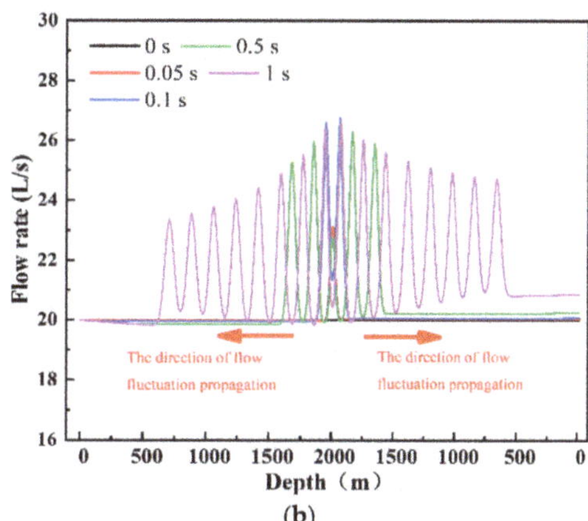

(b)

Figure 8. Wave propagation in the wellbore. (**a**) Pressure propagation; (**b**) Flow fluctuation propagation.

Figure 9 shows the pressure-flow fluctuation curve at the bottom hole and the tool. It is indicated that in the initial flow stage, the fluctuation of fluid flow is unstable. With time, the fluctuation of pressure and flow rate at the bottom and tool become stable. For a 2000 m vertical well, with an average flow rate of 20 L/s, the fluctuation of pressure and flow rate in the wellbore become stable after 50 s.

Figure 10 shows the fluctuations of pressure and flow rate at different times in the wellbore. It can be found that when fluctuation becomes stable in the wellbore, the flow rate at different positions fluctuates up and down near the initial flow rate. At the same time, the flow rate near the tool fluctuates more violently, and the flow rate fluctuates smoothly near the wellhead. Due to the limitation of the constant flow boundary condition at the drill pipe inlet, the flow rate is always maintained at 20 L/s at the inlet. From Figure 10b, it is found that the initial pressure in the drill pipe is the highest. As the pulse jet generator works stably, the flow area is always higher than that at the initial moment. The pressure drop of the pulse jet tool is always lower than that at the initial moment. This results in the situation mentioned above. In addition, comparing the pressure fluctuations in the drill pipe and the annulus, it is indicated that the pressure fluctuation in the drill pipe is slightly stronger than that in the annulus. Due to the constant pressure boundary condition at the annulus outlet, the outlet pressure is maintained as atmospheric pressure during calculation.

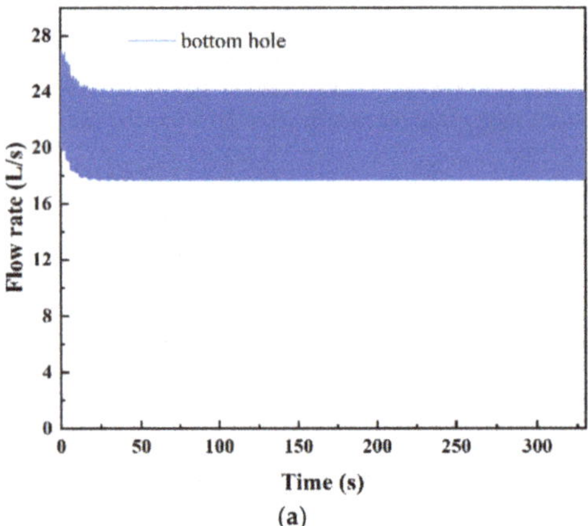

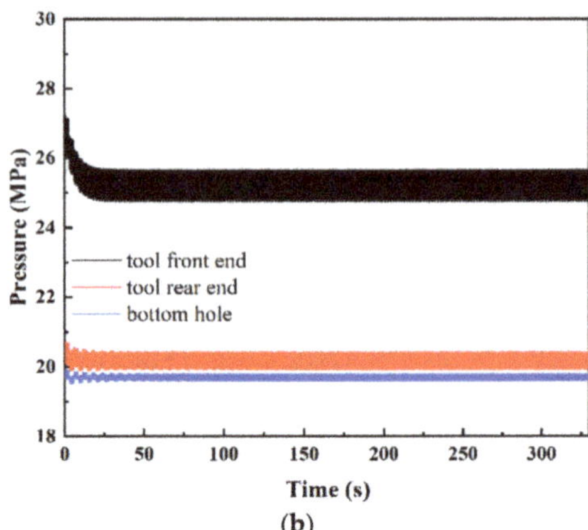

Figure 9. Fluctuation of flow rate and pressure at bottom hole and tool. (**a**) Flow rate at the bottom hole; (**b**) Pressure at bottom and tool.

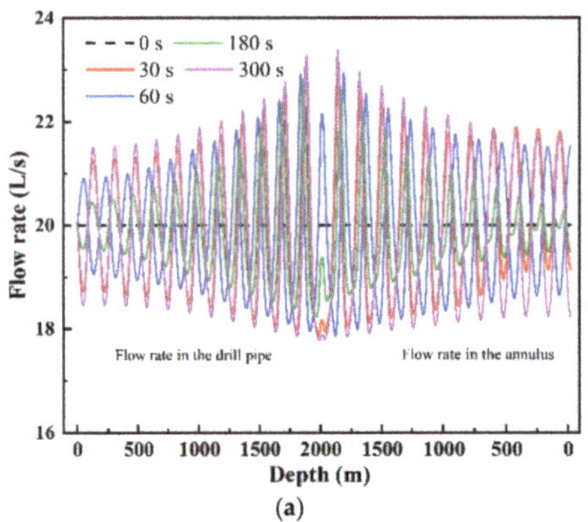

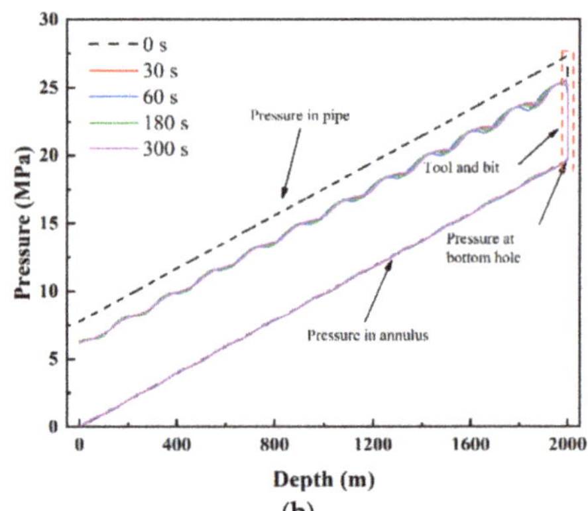

Figure 10. Fluctuation of pressure and flow rate in the wellbore at different times. (**a**) Flow rate fluctuation in the wellbore; (**b**) Pressure fluctuation in the wellbore.

4.2. Influence of Flow Rate

In order to obtain the displacement influence on pressure fluctuation and flow rate fluctuation at the bottom during drilling, cases in a 6000 m vertical well were carried out. The displacement range was 20 to 40 L/s. Figure 11 demonstrates the fluctuations of flow rate and pressure at the bottom hole and near the tool at different displacements. From Figure 11a, it is found that the fluid flow at the bottom hole fluctuates up and down near the corresponding displacement. The amplitude of fluctuation increases, but its fluctuation frequency does not change with increased displacement. Figure 11b indicates that the pressure fluctuation at the front end of the pulse jet generator is the most violent, followed by the rear end of the tool. When fluid passes through the bit nozzle, the pressure fluctuation decreases sharply. Comparing Figure 11a,b, the pressure fluctuation trend at the front end of the tool is opposite to the flow rate fluctuation trend. The maximum flow rate corresponds to the minimum pressure at the front end of the tool. The pressure fluctuation at the rear end of the tool is the same as its flow fluctuation. The reason is that the flow area changes with rotation of the impeller of the pulse jet tool. The smallest flow area corresponds to the minimum flow rate at the bottom hole. At the same time, the fluid tends to accumulate at the front end of the tool due to inertia, which causes the pressure to increase. Similarly, fluid inertia generates a suction at the rear end of the tool and the bottom hole, which causes the pressure to decrease.

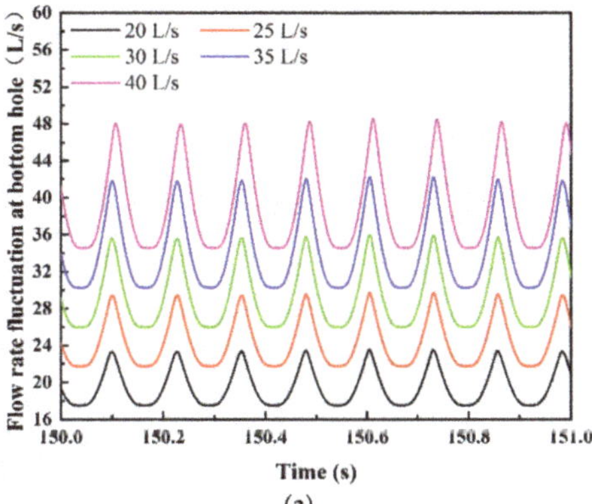

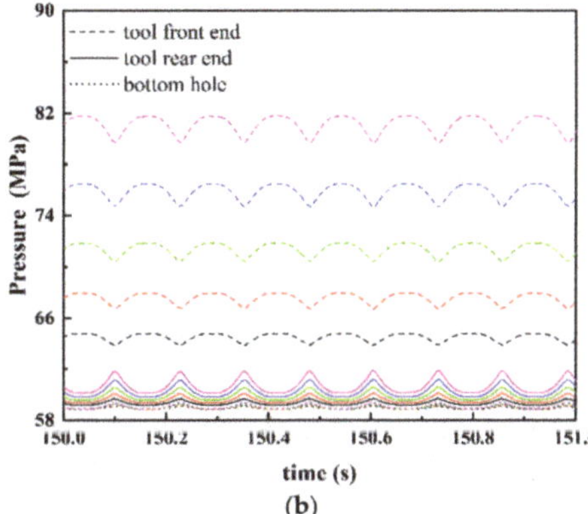

Figure 11. The fluctuations of flow rate and pressure at the bottom hole and near the tool at different displacements. (**a**) Flow rate fluctuation at the bottom hole; (**b**) Pressure fluctuation of bit and bottom hole.

In order to obtain the regularity of the fluctuation amplitude of flow rate and pressure with displacement, fluctuation amplitude is calculated, as shown in Figure 12. From the figure, it is found that the fluctuation at the tool is higher than that at the bottom hole. The fluctuation reaches a minimum after fluid flows through the nozzle. With flow displacement increasing from 20 L/s to 40 L/s, pressure fluctuation at the front end of the tool increases from 0.98 MPa to 2.18 MPa. The pressure fluctuation at the rear end changes from 0.49 MPa to 1.78 MPa. The pressure fluctuation at the bottom hole is enhanced from 0.23 MPa to 0.53 MPa. The flow rate fluctuation changes from 6.13 L/s to 14.32 L/s.

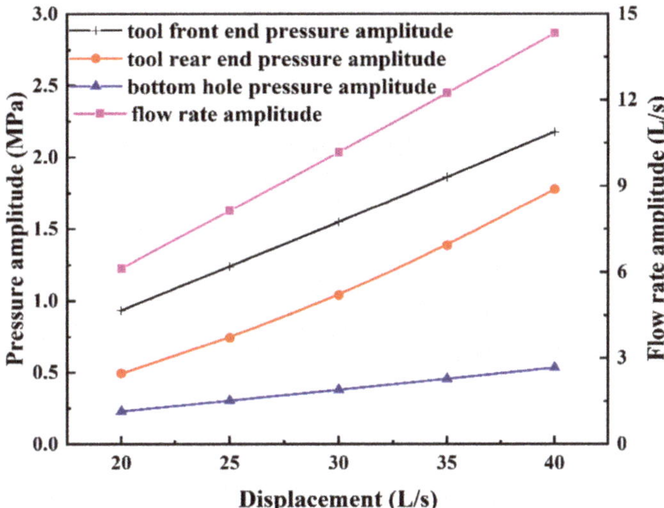

Figure 12. The amplitude of flow rate and pressure at different displacements.

4.3. Influence of Well Depth

Viscosity of 1 mPa s, with a flow rate of 20 L/s, was selected to analyze the influence of well depth on pulsed jet fluctuation. The well depth was from 2000 m to 10,000 m. From Figure 13, it is found that with depth increasing, the instantaneous flow rate at the bottom hole decreases slightly. The degree of reduction is most obvious when instantaneous flow rate reaches maximum and minimum values.

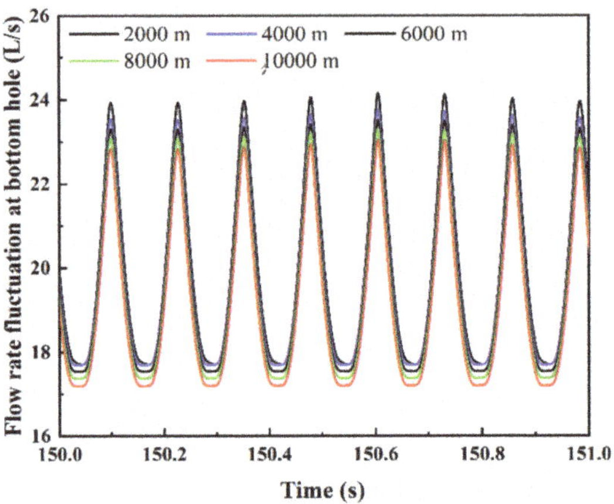

Figure 13. The fluctuations of flow rate at the bottom hole with different well depths.

Figure 14 shows the amplitude of flow rate and pressure with different well depths. With depth increasing, fluctuation amplitude of the flow rate decreases dramatically at first. When depth is over 4000 m, fluctuation amplitude of the flow rate decreases slightly with increase in well depth. Moreover, pressure fluctuation amplitude at the front end of the tool increases dramatically with increase of well depth from 2000 m to 4000 m. In contrast, amplitude decreases slightly when well depth is over 4000 m. Pressure amplitudes at the rear end of the tool and at the bottom hole remain stable with variation in well depth. The pressure amplitude at the bottom hole is about 0.23 MPa. The flow rate amplitude decreases from 6.6 L/s to 6.0 L/s when well depth increases from 2000 m to 10,000 m.

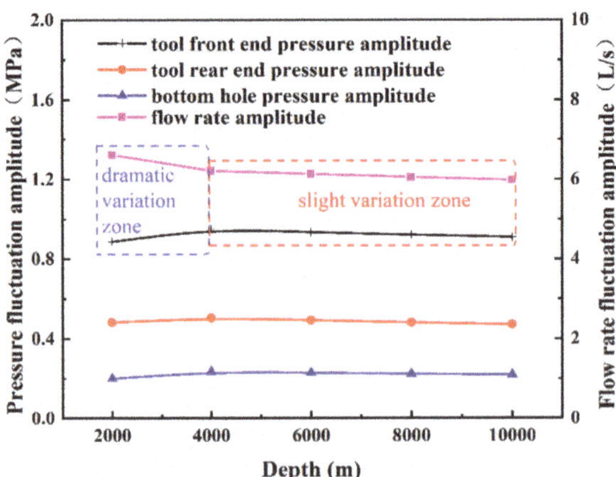

Figure 14. The range of pressure and flow rate fluctuation at the bottom hole and tool.

4.4. Influence of Drilling Fluid Viscosity

The amplitude of flow rate and pressure fluctuation with different viscosities is shown in Figure 15. In these cases, the well depth was 6000 m, and the displacement was 20 L/s. The other parameters were the same as in Table 2. It can be seen from the figure that with increase of drilling fluid viscosity, fluctuation amplitude of pressure and flow at the tool and the bottom hole remain stable without obvious change. The pressure amplitude at the bottom hole is about 0.23 MPa. The flow rate amplitude is about 6.15 L/s. The results indicate that drilling fluid viscosity has an extremely slight effect on amplitude of pressure and flow rate at the bottom hole under pulsed jet.

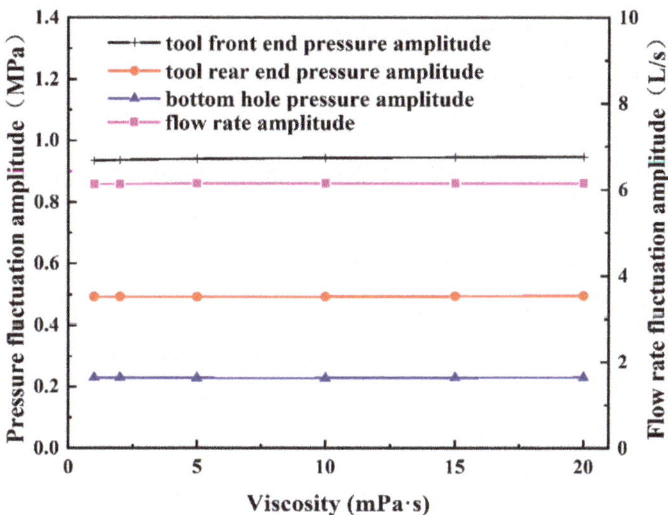

Figure 15. The amplitude of flow rate and pressure with different viscosities.

Table 2. The basic parameters of the wellbore, drill pipe, and drill fluid.

Parameters	Values	Parameters	Values
Drill pipe inner diameter, mm	108.62	Casing elastic modulus, Pa	2.0×10^{11}
Drill pipe outer diameter, mm	127	Casing Poisson's ratio	0.29
Drill pipe elastic modulus, Pa	2.0×10^{11}	Drilling fluid density, kg/m^3	1000
Poisson's ratio of drill pipe	0.3	Elastic modulus of drill fluid, Pa	2.2×10^{9}
Casing inner diameter, mm	257.18	The size of the nozzle, cm	1.4 cm × 5
Drill pipe inner diameter, mm	108.62	Casing elastic modulus, Pa	2.0×10^{11}
Drill pipe outer diameter, mm	127	Casing Poisson's ratio	0.29
Drill pipe elastic modulus, Pa	2.0×10^{11}	Drilling fluid density, kg/m^3	1000
Poisson's ratio of drill pipe	0.3	Elastic modulus of drill fluid, Pa	2.2×10^{9}

4.5. Influence of Flow Area in the Tool

The change of the flow area in the pulse jet generator affects drop in tool pressure at different instantaneous flow rates. Simulations carried out about the flow area of the pulse jet generator changed from 2 cm^2 to 8 cm^2 in a 6000 m well. The displacement and viscosity were 20 L/s and 1 mPa·s, respectively. The other parameters were the same as in Table 2.

In order to explore the influence of tool flow area on instantaneous bottom hole pressure and flow rate fluctuations, the fluctuations of the pressure and flow rate were extracted, as shown in Figures 16 and 17. By comparing the pressure fluctuations near the tool and at the bottom hole under different tool flow area conditions, it is found that when the minimum flow area of the tool is 2 cm^2, the fluctuations of the pressure and flow rate are the most severe. When the flow area is over 6 cm^2, the flow rate becomes slightly lower than 20 L/s.

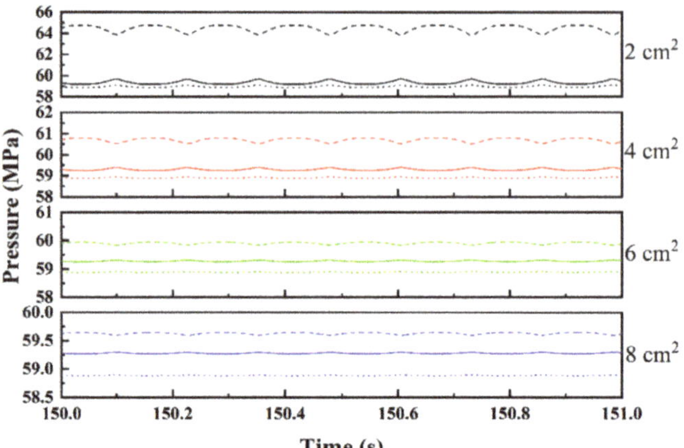

Figure 16. The fluctuations of pressure at the bottom hole and near tools with different flow areas of the tool.

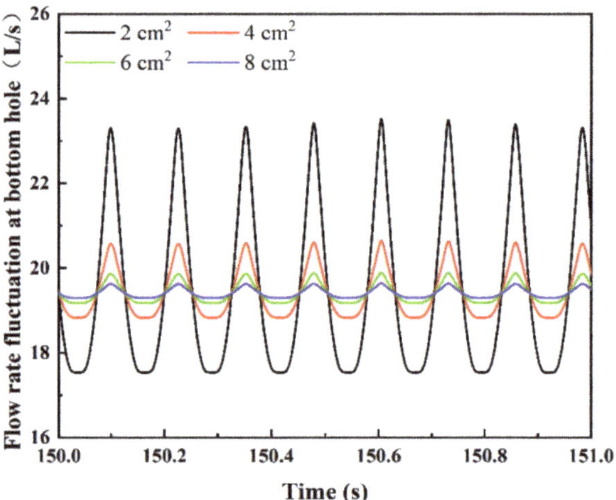

Figure 17. The fluctuations of flow rate at the bottom hole and near tools with different flow areas of the tool.

Figure 18 shows the amplitude of flow rate and pressure with different flow areas of the tool. From the statistics, it can be found that tool flow area greatly influences its fluctuation amplitude. When the flow area increased from 2 cm^2 to 4 cm^2, the amplitude of flow rate fluctuation decreased by 70.5%, from 4.0 L/s to 1.1 L/s. The bottom hole pressure fluctuation amplitude decreased by 69.6%, from 0.23 MPa to 0.07 MPa. The results indicate that the pressure and flow rate fluctuations are greatly affected by the minimum flow area in the pulse jet generator. When the minimum flow area of the tool increases, the fluctuations of the pressure and flow rate decrease rapidly in the first stage. With further increase of the minimum flow area, the amplitudes of pressure and flow rate decrease slightly.

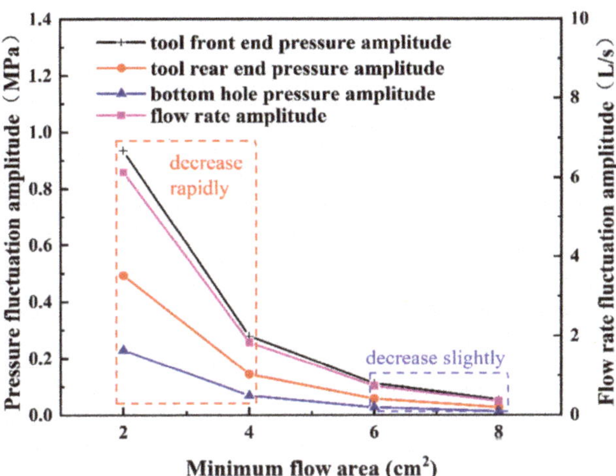

Figure 18. The amplitude of flow rate and pressure with different flow areas of the tool.

4.6. Cuttings Mechanical Characteristics Analysis at the Bottom Hole under the Pulsed Jet

Force acted on cuttings at the bottom hole under pulsed jet and continuous jet are compared. Table 3 shows the basic parameters of the cases.

Table 3. The basic parameter of a single-particle model of cuttings at the bottom hole.

Parameters	Values	Parameters	Values
Cuttings diameter, mm	2	Nozzle diameter, mm	14
Cuttings density, kg/m^3	2700	Distance from the nozzle to bottom, mm	30
Pore pressure, MPa	58.86	Nozzle angle, °	20
Fracture width at the bottom hole, mm	0.5	Fluid density, kg/m^3	1000

The final state of the cuttings can be obtained according to the component forces F_1 and F_2. Figures 19 and 20 are the component force states of the cuttings under continuous jet and pulsed jet, respectively. Moreover, the corresponding instantaneous flow rate and pressure at the bottom hole are shown in the figures. From Figure 19, it is indicated that pulsed jet makes the component force F_1 fluctuate compared with continuous jet. When the value of F_1 is positive, it means that the cuttings will tumble away from the bottom hole. As seen in the figure, the instantaneous component force F_1 becomes positive, corresponding to maximum instantaneous flow rate and pressure. At this time, the cuttings will tumble away from the bottom hole. Component force F_2 is also analyzed in Figure 20. It is found that the minimum state of component force F_2 corresponds to the minimum instantaneous flow rate. In all cases simulated in this paper, the value of F_2 is always positive. By combining the results in the figures, it can be concluded that the pulsed jet generates the flow rate and pressure fluctuation at the bottom hole. Fluctuation improves the force acted on the cuttings and helps the cuttings tumble off the bottom hole.

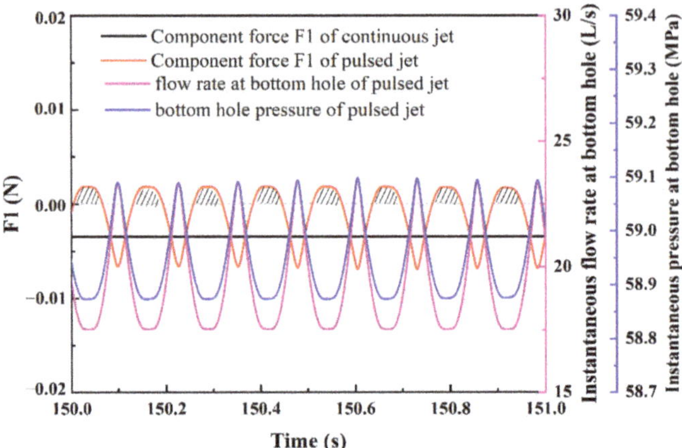

Figure 19. Comparison of particle force F_1 at the bottom hole with continuous jet and pulse jet.

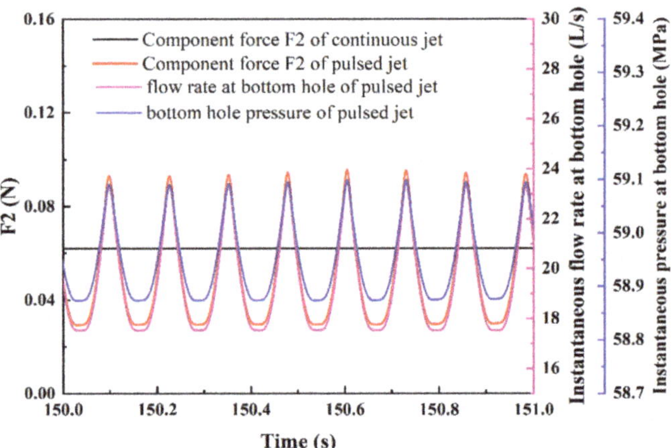

Figure 20. Comparison of particle force F_2 at the bottom hole with continuous jet and pulse jet.

In addition, all forces of the cuttings are calculated under the pulsed jet condition, as shown in Figure 21. Due to the flow fluctuation at the bottom hole, the pressure force of the drilling fluid column also fluctuates. Therefore, the pulsed jet can reduce the chip hold-down effect at the bottom hole. Moreover, the flow rate fluctuation caused by the pulsed jet generator induces frequent variation in relative velocity between fluid and cuttings. The pulsed jet causes fluctuations of the particle Reynolds number, drag coefficient, and drag force. Fluctuated drag force is beneficial for cuttings to break away from the bottom hole.

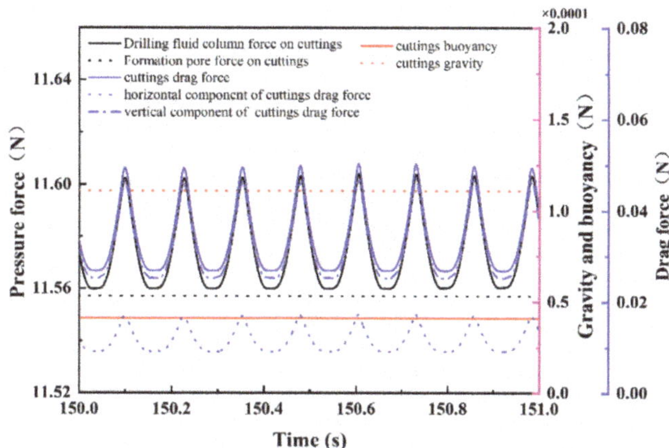

Figure 21. Force analysis of cuttings at the bottom hole under the pulsed jet.

5. Conclusions

Pulsed jet drilling can generate the fluctuation of pressure and flow rate at the bottom hole, which is beneficial for cuttings cleaning. It is expected to improve the ROP during drilling in unconventional oil and gas formations. In this paper, a pressure-flow rate fluctuation model is established. Based on the model, the force acted on the cuttings is analyzed to reveal the mechanism of cuttings cleaning under pulsed jet at the bottom hole. The main conclusions are as follows:

(1) For a 2000 m well, the pulsed jet can generate pressure fluctuation at the bottom hole from 0.23 MPa to 0.53 MPa, corresponding to flow rate amplitude from 6.13 L/s to 14.32 L/s with displacement range of 20 L/s to 40 L/s.

(2) The flow rate amplitude at the bottom hole decreases dramatically with increase in well depth when well depth is less than 4000. When well depth is over 4000 m, the flow rate amplitude varies slightly with variation of well depth. Pressure amplitude at the bottom hole keeps stable with increase in well depth.

(3) Fluid viscosity has a slight influence on flow fluctuation at the bottom hole. The amplitude of pressure and flow rate fluctuation is mainly affected by the flow area of the tool. When the flow area increased from 2 cm^2 to 4 cm^2, the amplitude of flow fluctuation decreased by 70.5%, and the amplitude of pressure decreased by more than 60%.

(4) Pulsed jet generates fluctuation in the instantaneous drilling fluid column force acted on the cuttings and reduces the chip hold-down effect. In addition, the pulsed jet caused fluctuation of drag force to assist the cuttings in breaking away from the bottom hole.

Author Contributions: Conceptualization, H.Z., H.S. and Z.H.; methodology, H.Z.; formal analysis, H.Z. and Z.C.; investigation, H.Z. and Z.G.; resources, H.Z., H.S. and Z.H.; data curation, H.Z.; writing—original draft preparation, H.Z.; writing—review and editing, H.Z. and F.G.; visualization, H.Z. and F.G.; supervision, H.Z., H.S. and Z.H.; project administration, H.S. and Z.H.; funding acquisition, H.S. and Z.H. All authors have read and agreed to the published version of the manuscript.

Funding: This study was supported by the financial support of The Joint Funds of the National Natural Science Foundation of China (Grant No. U19B6003-05), National Key Research and Development Program of China (No. 2019YFA0708302), the Strategic Cooperation Technology Projects of CNPC and CUPB (ZLZX2020-01).

Institutional Review Board Statement: Not applicable.

Informed Consent Statement: Not applicable.

Data Availability Statement: Not applicable.

Conflicts of Interest: The authors declare no conflict of interest.

Nomenclature

v	Fluid velocity, m/s
ρ	Density of the fluid, kg/cm^3
τ_p	Shear stress on the inner wall of the drill pipe, Pa
A_p	Flow area of the drill pipe, m^2
α	Compressibility coefficient of the fluid, 1/Pa
β	Expansion coefficient of the flow channel, 1/Pa
C	Propagation velocity of the pressure wave, m/s
f_a	Darcy-Weisbach friction factor
D	Inner diameter of the drill pipe, m
k	Brunone friction coefficient
C^*	Vardy shear attenuation coefficient
d_p	Diameter of the cuttings, m
ρ_p,ρ_l	Density of cuttings and fluid, kg/m^3
τ_r	Relaxation time in the drag force model
μ	Viscosity of the fluid, Pa·s
C_D	drag coefficient
Re_p	particle Reynolds number
G	Gravity of the cuttings, N
F_b	Buoyancy of the cuttings, N
F_c	Pressure force of the fluid column, N
F_D	Drag force, N
F_p	Pressure force of the pore pressure, N
F_{Dh}, F_{Dv}	Horizontal and vertical components of the drag force, N
F_N	Support force acted on the cuttings, N
F_f	Friction force acted on the cuttings, N
a_1, a_2, a_3	The coefficient in the Morsi and Alexander drag model
$\vec{u}$, $\vec{u}_p$	Velocity vector of the fluid and cuttings, m/s

References

1. Barton, S.; Baez, F.; Alali, A. Drilling performance improvements in gas shale plays using a novel drilling agitator device. In Proceedings of the North American Unconventional Gas Conference and Exhibition, The Woodlands, TX, USA, 14–16 June 2011; OnePetro: Richardson, TX, USA, 2011.
2. Baez, F.; Barton, S. Delivering performance in shale gas plays: Innovative technology solutions. In Proceedings of the SPE/IADC Drilling Conference and Exhibition, Amsterdam, The Netherlands, 1–3 March 2011; OnePetro: Richardson, TX, USA, 2011.
3. Fry, M. Urban gas drilling and distance ordinances in the Texas Barnett Shale. *Energy Policy* **2013**, *62*, 79–89. [CrossRef]
4. Maurer, W.C.; Rowley, J.; Carwile, C. *Advanced Turbodrills for Geothermal Wells*; Maurer Engineering Inc.: Houston, TX, USA; Los Alamos Scientific Lab.; Los Alamos, NM, USA; Department of Energy: Washington, DC, USA; Div. of Geothermal Energy: Washington, DC, USA, 1978.
5. Conn, A. On the fluid dynamics of working water jets: Continuous, pulsed and cavitating. In Proceedings of the 5th Pacific Rim International Conference on Water Jet Technology, New Dehli, India, 3–5 February 1998.
6. Ghalambor, A.; Hayatdavoudi, A.; Akgun, F.; Okoye, C. Intermittent nozzle fluid flow and its application to drilling. *J. Pet. Technol.* **1988**, *40*, 1021–1027. [CrossRef]
7. Bizanti, M.S. Jet pulsing may allow better hole cleaning. *Oil Gas J.* **1990**, *88*, 67–68.
8. Kolle, J. *HydroPulse Drilling*; Tempress Technologies, Inc.: Kent, WA, USA, 2004.
9. Walter, B.H. Flow Pulsing Method and Apparatus for the Increase of the Rate of Drilling. Google Patents 6053261, 25 April 2000.
10. Walter, B.H. Flow Pulsing Apparatus for Use in Drill String. US Patent 4819745, 4 November 1989.
11. Li, G.; Shi, H.; Liao, H.; Shen, Z.; Niu, J.; Huang, Z.; Luo, H. Hydraulic pulsed cavitating jet-assisted drilling. *Pet. Sci. Technol.* **2009**, *27*, 197–207. [CrossRef]
12. Fu, J.; Li, G.; Shi, H.; Niu, J.; Huang, Z. A novel tool to improve the rate of penetration—Hydraulic-pulsed cavitating-jet generator. *SPE Drill. Complet.* **2012**, *27*, 355–362. [CrossRef]
13. Li, G.; Shi, H.; Niu, J.; Huang, Z.; Tian, S.; Song, X. Hydraulic pulsed cavitating jet assisted deep drilling: An approach to improve rate of penetration. In Proceedings of the International Oil and Gas Conference and Exhibition in China, Beijing, China, 8–10 June 2010; OnePetro: Richardson, TX, USA, 2010.

14. Liu, Y.; Wei, J.; Ren, T. Analysis of the stress wave effect during rock breakage by pulsating jets. *Rock Mech. Rock Eng.* **2016**, *49*, 503–514. [CrossRef]
15. Dehkhoda, S. Experimental and Numerical Study of Rock Breakage by Pulsed Water Jets. Ph.D. Thesis, University of Queensland, Brisbane, Australia, 2011.
16. Zelenak, M.; Foldyna, J.; Scucka, J.; Hloch, S.; Riha, Z. Visualisation and measurement of high-speed pulsating and continuous water jets. *Measurement* **2015**, *72*, 1–8. [CrossRef]
17. Zhang, W.; Shi, H.; Li, G.; Song, X. Fluid hammer analysis with unsteady flow friction model in coiled tubing drilling. *J. Pet. Sci. Eng.* **2018**, *167*, 168–179. [CrossRef]
18. Zhang, W.; Shi, H.; Li, G.; Song, X.; Zhao, H. Mechanism analysis of friction reduction in coiled tubing drilling with axial vibratory tool. *J. Pet. Sci. Eng.* **2019**, *175*, 324–337. [CrossRef]
19. Ahn, S.; Horne, R.N. The use of attenuation and phase shift to estimate permeability distributions from pulse tests. In Proceedings of the SPE Annual Technical Conference and Exhibition, Denver, CO, USA, 30 October–2 November 2011; OnePetro: Richardson, TX, USA, 2011.
20. Rosa, A.; Horne, R. Reservoir description by well test analysis using cyclic flow rate variation. *SPE Form. Eval.* **1997**, *12*, 247–254. [CrossRef]
21. Fokker, P.A.; Borello, E.S.; Verga, F.; Viberti, D. Harmonic pulse testing for well performance monitoring. *J. Pet. Sci. Eng.* **2018**, *162*, 446–459. [CrossRef]
22. Bergant, A.; Ross Simpson, A.; Vítkovsk, J. Developments in unsteady pipe flow friction modelling. *J. Hydraul. Res.* **2001**, *39*, 249–257. [CrossRef]
23. Vardy, A.E.; Brown, J.M.B. *Turbulent, Unsteady, Smooth-Pipe Friction*; BHR Group Conference Series Publication: Milton Koynes, UK; Mechanical Engineering Publications Limited: London, UK, 1996; pp. 289–312.
24. Saadati, M.; Forquin, P.; Weddfelt, K.; Larsson, P.-L.; Hild, F. Granite rock fragmentation at percussive drilling—Experimental and numerical investigation. *Int. J. Numer. Anal. Methods Geomech.* **2014**, *38*, 828–843. [CrossRef]
25. Saksala, T.; Gomon, D.; Hokka, M.; Kuokkala, V.-T. Numerical and experimental study of percussive drilling with a triple-button bit on Kuru granite. *Int. J. Impact Eng.* **2014**, *72*, 56–66. [CrossRef]
26. Wang, R.-H.; Du, Y.-K.; Ni, H.-J.; Lin, M. Hydrodynamic analysis of suck-in pulsed jet in well drilling. *J. Hydrodyn. Ser. B* **2011**, *23*, 34–41. [CrossRef]
27. Gosman, A.; Loannides, E. Aspects of computer simulation of liquid-fueled combustors. *J. Energy* **1983**, *7*, 482–490. [CrossRef]
28. Morsi, S.A.; Alexander, A.J. An investigation of particle trajectories in two-phase flow systems. *J. Fluid Mech.* **1972**, *55*, 193–208. [CrossRef]
29. Zhang, W.; Chen, X.; Pan, W.; Xu, J. Numerical simulation of wake structure and particle entrainment behavior during a single bubble ascent in liquid-solid system. *Chem. Eng. Sci.* **2022**, *253*, 117573. [CrossRef]
30. Peterseim, D.; Schedensack, M. Relaxing the CFL condition for the wave equation on adaptive meshes. *J. Sci. Comput.* **2017**, *72*, 1196–1213. [CrossRef]

Article

An NMR Investigation of the Influence of Cation Content in Polymer Ion Retarder on Hydration of Oil Well Cement

Zhigang Qi [1,*], Yang Chen [1], Haibo Yang [1], Hui Gao [1], Chenhui Hu [2] and Qing You [2]

[1] Drilling Technology Research Institute, Sinopec Shengli Petroleum Engineering Co., Ltd., Dongying 257017, China
[2] School of Energy Resources, China University of Geosciences, Beijing 100083, China
* Correspondence: qizhg25.ossl@sinopec.com

Abstract: Low field pulse nuclear magnetic resonance (LF-NMR) was used to analyze the effects of a polymeric ion retarder and the amount of acryloxyethyl trimethylammonium chloride (DMC) in the retarder on the distribution of T_2, thickening property, and strength of cement paste. The effect of pressure and temperature on the thickening curve was investigated, and the hydration products were analyzed using XRD. The result shows that the wrapped water of the precipitation is the main reaction aqueous phase of cement slurry in the hydration, with short T_2 time and a large relaxation peak area. The retarder weakens the van der Waals force and electrostatic adsorption force between the water and cement particles, reducing the hydration rate of cement particles. An appropriate increase in the cationic content of polymeric ion retarder can improve the early strength of cement slurry.

Keywords: NMR; oil well cement; hydration; retarder

Citation: Qi, Z.; Chen, Y.; Yang, H.; Gao, H.; Hu, C.; You, Q. An NMR Investigation of the Influence of Cation Content in Polymer Ion Retarder on Hydration of Oil Well Cement. *Energies* **2022**, *15*, 8881. https://doi.org/10.3390/en15238881

Academic Editor: Daoyi Zhu

Received: 20 September 2022
Accepted: 21 November 2022
Published: 24 November 2022

1. Introduction

In order to improve crude oil production and support economic development, oil exploration and development has begun to expand to deep oil and gas reservoirs. In particular, the discovery of a large number of ultra-deep wells in recent years has brought huge oil reserves and great challenges in cementing operations [1–3]. A tremendous amount of cementing practices have shown that the technical difficulty of the cementing operation in the long well section is that the large temperature difference formed in the long well section causes the top cement slurry to be over retarded, and the key to cause the problem is the retarder. Due to the large temperature difference (>40 °C) between the top and bottom of the well, the cement additives are used across the temperature zone, seriously reducing the use effect of most of the additives and cementing quality [4]. In order to ensure the safe pumping of cement slurry under high temperature conditions, high-temperature retarders are usually introduced into cement slurry. Most of the high-temperature retarders currently cannot adapt to the large temperature difference. When the cement slurry returns from the bottom to the top low-temperature well section, the retarding effect cannot be removed in time, and the strength of the cement slurry develops more slowly at low temperature, which make it difficult to guarantee the quality of the top cementing and ensure the safety of the well [5].

Water involved in the hydration process of the oil well cement can be classified into three types, which are chemically bound water, physically bound water, and free water. The three types of water undergo a dynamic transition during the hydration process. Cement additives can improve the performance of cement slurry and can have a significant impact on the hydration reaction of cement. For example, the retarder can prolong the thickening time and ensure the safe operation of the cement injection process in the deep well with a large temperature difference and long sealing section. Currently, the retarders used in the deep well with large temperature difference and long sealing section are always based on the 2-acrylamide-2-methylpropanesulfonic acid (AMPS). 2-2-Acrylamide-2-methylpropanesulfonic acid (AMPS) is one of the most common copolymerization main monomers, which consists of oil well cement retarders at home and abroad. Its structure is characterized by sulfonic acid groups, amide groups, high charged carbonyl group and the large side chain, which makes AMPS resistant to high temperature and salt, acid and alkali, with excellent thermal stability, adsorption, complexation and hydrolysis stability [6]. Liang et al. [7] synthesized polymer retarder PAIN using AMPS, itaconic acid (IA) and N-vinylpyrrolidone (NVP); regulator isopropanol and redox system are initiators during the synthesis process, and its five-membered pyrrole ring structure makes the cement paste system with PAIN obtain high-temperature stability and a thickening time of 256 min at 110 °C and 234 min at 130 °C. In order to overcome the problems of over-sensitivity, PC-H42L was synthesized by Hu et al. [8] using AMPS and two unsaturated carboxylic acid monomers, and the thickening time was controlled in 4–5 h by adjusting the retarder dosage in the range of 60–210 °C. It has no adverse effect on the stability of cement paste settlement and water loss, and also has good anti-flutter performance. Guo et al. [9] synthesized polymeric retarders under acidic conditions with AMPS and IA at 60 °C. For this kind of retarder, the suitable temperature ranges from 50 °C to 180 °C; it can significantly extend the thickening time of cement slurry, and exert an excellent high-temperature retarding effect. However, with retarders, if not controlled properly, the cementing safety and cementing quality of the low-temperature section of the well cannot be guaranteed [10–12]. Therefore, exploring the influence of retarder on different types of water during cement hydration reaction has important significance for better understanding the hardening mechanism of cement, guiding the optimization of the cementing slurry system and ensuring the safety of the cementing operation [13,14].

With the development of the nuclear magnetic resonance (NMR) technology, NMR has been applied as a non-destructive testing technique to characterize the pore structures of cement-based materials [15,16]. Nuclear magnetic resonance analysis technology can be roughly divided into three parts: magnetic resonance imaging (MRI), nuclear magnetic resonance spectroscopy, and time domain nuclear magnetic resonance (TD-NMR) technology [17–20]. These three parts are commonly referred to as low field nuclear magnetic resonance technology (LF-NMR). LF-NMR mainly uses hydrogen protons as probes to study the mobility of molecules. In the analysis of LF-NMR, the transverse relaxation time T_2 is often used to reflect the movement of molecules by measuring the relaxation time of the molecules to characterize the pore structure changes, swelling and phase transition in the sample [21]. In the cement slurry hydration process, the existence form of the water molecule can be judged by the different relaxation time of the hydrogen atom.

In this paper, low field pulse nuclear magnetic resonance (LF-NMR) was used to analyze the relaxation signal of protons in water molecules to study the mechanism of water and polymeric ion retarders in the cement-hardening process. The process by which a nuclear spin system in a high-energy state transfers energy to a similar magnetic nucleus in a neighboring low-energy state is called spin–spin relaxation, or transverse relaxation [22,23]. This process is only the energy exchange of the spin state of similar magnetic nuclei, and does not cause the change of the total nuclear magnetic energy. Its half-life is expressed in T_2. In particular, the T_2 distribution of cement slurry was analyzed to determine the existing state of water and the retarding mechanism at the early stage of the cement-hardening reaction [24,25]. The T_2 in this paper was acquired with the standard Carr–Purcell–Meiboom–Gill (CPMG) sequence from the built-in software.

This article aims to explore the hardening mechanism of cement slurry and better synthetic conditions of the retarder suitable for a large temperature difference, which has positive significance for building up a kind of cement slurry system suitable for the development of oil and geothermal resources in deep formations.

2. Materials and Methods

2.1. Reagents and Instruments

2-Acryloylamino-2-methyl-1-propanesulfonic acid (AMPS), itaconic acid (IA), methoxy-polyethylene glycols (MG) NaOH, and $K_2S_2O_8$ are all analytical reagents (ARs) and were purchased from Shanghai Jizhi Biochemical Technology Co., Ltd. (Shanghai, China). Methacryloxyethyltrimethyl ammonium chloride (DMC) is an AR from Wuhan Fengtai Weiyuan Technology Co., Ltd. (Wuhan, China). Shengwei G-grade oil well cement was provided by the Scientific Research Institute of Shengli Oilfield cementing center. Defoamers were provided by the Scientific Research Institute of well cementing center of Shengli Oilfield; fluid loss additive was self-made by the laboratory.

The MacroMR movable fluid distribution detection nuclear magnetic resonance imager was purchased from Suzhou Newmark Electronic Technology Co., Ltd. (Suzhou, China). The X'PERT Pro X-ray diffractometer was purchased from Panaco (Amsterdam, the Netherlands). The HT/HP thickener was purchased from Chandler company (Santa Ana, CA, USA). The HTD7370 HT/HP curing kettle was purchased from Qingdao Haitongda Special Instrument Co., Ltd. (Qingdao, China), and the Yaw-300 pressure testing machine was purchased from Jinan Liling Testing Machine Co., Ltd. (Jinan, China).

The cement slurry formula in this paper was as follows: G-grade oil well cement + IADN polymeric ion retarder (0, 0.5%) + fluid loss additive (0.8%) + defoamer (3–5 drops). The water–cement ratio was 0.44, the curing temperature was 85 °C, and the curing duration were 0.5 h, 1 h, 1.5 h, 2 h, 2.5 h, and 3 h, respectively.

2.2. Synthesis of IADN Retarder

A solution of 2 mol% IA, 3 mol% AMPS, 0.1 mol% DMC, and 0.03 mol% MG was first prepared. Then, NaOH was used to adjust the solution pH to 7–9, and 1.5 mol% of $K_2S_2O_8$ was added as the initiator. The mixture was put in a 65 °C water bath under nitrogen atmosphere for 5 h to create a viscous copolymer, which was then precipitated and purified with 100% ethanol, washed with acetone, and dried under vacuum at 40 °C for 1 day to obtain the final product of polymerized ionic retarder (IADN).

We changed the content of DMC from 0.1 mol% to 0 mol%, 0.05 mol%, 0.15 mol%, 0.25 mol%, 0.56 mol%, 0.6 mol%, 1.25 mol% in order to prepare three kinds of solutions. Then, we followed the above steps for the synthesis of IADN with 0%, 1%, 3%, 5%, 10%, 15%, and 20% DMC.

2.3. Analysis Methods

The LF-NMR was applied to measure the relaxation time (T_2) of the cement slurry during the hydration process to obtain the distribution, connectivity, and physical parameters of pores in the cement slurry in the form of a gel [26–28].

The hard pulse CPMG sequence was used to obtain the T_2 value of the cement slurry. The CPMG sequence, which is based on the spin echo pulse sequence, generates a series of 180-degree pulses to obtain the echo pulse sequence of multiple echo signals, and can improve the detection signal's signal-to-noise ratio. The attenuated echo signal reflects the relaxation signal of H^1 in the water; the T_2 distribution was calculated by Laplace transform or Fourier transform of CPMG echo signal of proton H^1. The relative quantities of chemically bound water, physically bound water, and free water in the cement slurry could be determined by computing the area under the T_2 distribution. The peak location and change in peak area of T_2 distribution represent the dynamic transformation of the states of water, providing an analytical method for studying the hydration process and curing mechanism of cement slurry [29].

A MacroMR nuclear magnetic resonance analyzer was employed for the LF-NMR. The permanent magnet's magnetic field intensity was (0.5 ± 0.08) t, the magnetic field frequency was 21.3 MHz, and the diameter of the probe coil was 60 mm. The number of echoes and the echo length were set to 1000 and 1200 ms, respectively. The T_2 distribution was plotted using the MacroMR moveable fluid moisture distribution detecting nuclear magnetic resonance imager.

The HT/HP thickener Yaw-300 pressure testing machine and X'PERT Pro X-ray diffractometer were used to obtain the thickening time, compressive strength and XRD pattern of the cement slurry. The HTD7370 HT/HP curing kettle was employed for curing the cement slurry.

3. Results and Discussion

3.1. Effect of Retarder on T_2 Distribution of Cement Slurry

As mentioned earlier, the relaxation time (T_2) of the cement slurry was obtained using the MacroMR nuclear magnetic resonance analyzer. Six T_2 distributions with different curing time were obtained for each group, and the results of the slurry samples without and with retarders are shown in Figures 1 and 2.

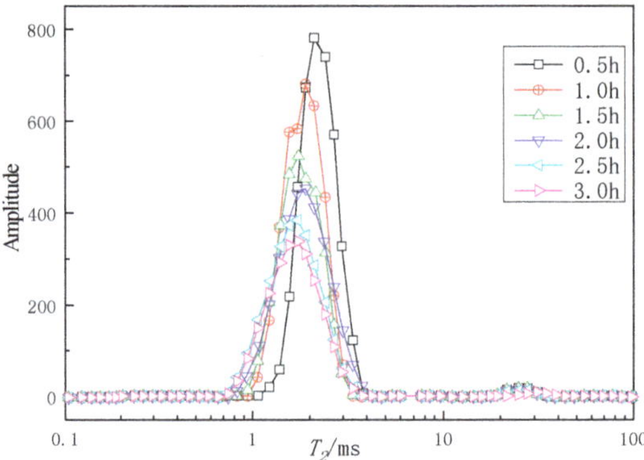

Figure 1. T_2 distribution of blank cement slurry.

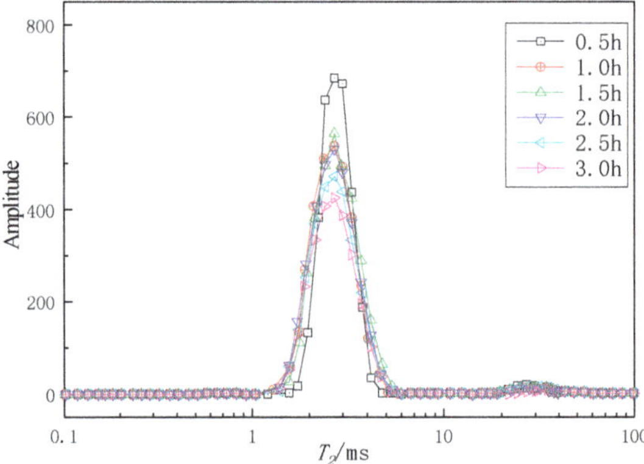

Figure 2. T_2 distribution of cement slurry with the retarder.

The T_2 distributions of the blank cement slurry and the cement slurry with retarder exhibit two distinctive peaks, as shown in Figures 1 and 2. The position of the peaks represents the different states of water in the slurry, namely, the peaks at 2 ms and 30 ms in the figures indicate that the slurry contained two phases of water. Comparing Figures 1 and 2, the peaks generally shifted to the right in the presence of the retarder. The area under the strong peak was significantly larger than that under the weak peak, showing the dominance of the type of the water phase represented by the former. However, as the curing time increased, the peak value of the strong peak rapidly dropped. This suggests that the relative amount of this type of aqueous phase declined rapidly during the hydration process, which was mirrored in the dramatic reduction in the corresponding peak area. The change in the peak (2 ms) area was greatly affected by the hydration reaction. For the peak at 30 ms, the reduction became less significant with the increase in the curing time.

At the early stage of hydration, C_3S and C_2S reacted with water to produce a large amount of hydrated calcium silicate (C-S-H) gel, as well as Ca^{2+}, OH^-, and $H_3SiO_4{}^-$ ions (Equation (1)). The hydration products then gathered and formed precipitates due to the van der Waals and electrostatic interactions between these ions. The water wrapped in the flocculation has the smaller relaxation time due to the high volume fraction of flocculation, while the free water in the cement slurry has the larger relaxation time. Thus, the peak at 30 ms corresponds to the relaxation signal of the wrapped water gathered in the flocculation, which is consistent with the relevant literature [30–33]:

$$2Ca_3SiO_5 + 8H_2O \rightarrow 6Ca^{2+} + 10OH^- + 2H_3SiO_4{}^- \quad (1)$$

Comparing Figures 1 and 2, the strong and weak peaks of cement slurry shifted to the right in the presence of the retarder, and the peak area at 30 ms was relatively small. This means that the retarder weakened the van der Waals force and electrostatic force on the water wrapped in the flocculant, reducing the extent of hydration reaction between cement particles and water, resulting in the reduction in the amount of C-S-H gel generated and the initial strength of the cement, and the increase in the hardening time.

3.2. Effect of DMC Content in Retarder on Relaxation Time T_2

As described earlier, the retarder was prepared under three synthesizing conditions with DMC contents of 0%, 10%, and 20%, respectively. The T_2 distributions of the cement slurry after curing for 0.5 h, 5 h, and 12 h are shown in Figures 3–5.

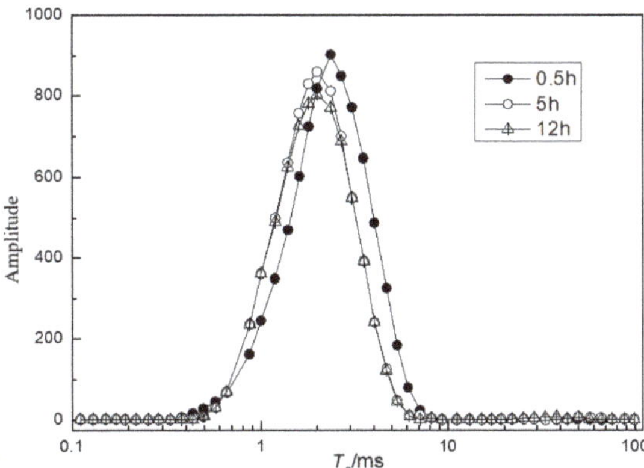

Figure 3. T_2 distribution of cement slurry with DMC content of 0%.

It can be seen from the figures that the T_2 peak of the cement slurry shifted to the left, and the peak value dropped as the curing time increased. Moreover, the peak value of cement slurry containing the DMC cationic retarder was significantly higher in the early hydration stage (within the first 5 h). To investigate the corresponding relationship between T_2 distribution and retarder synthesized with different contents of DMC, the relative content of amorphous water in the hydration products was evaluated using the peak area. The area under the strong peak and the weak peak can respectively indicate the relative amount of water wrapped in the flocculation and free water in the cement slurry. Figure 6 shows the change in the peak area with the cement slurry curing time to demonstrate the effect of different retarders on the hydration process of cement.

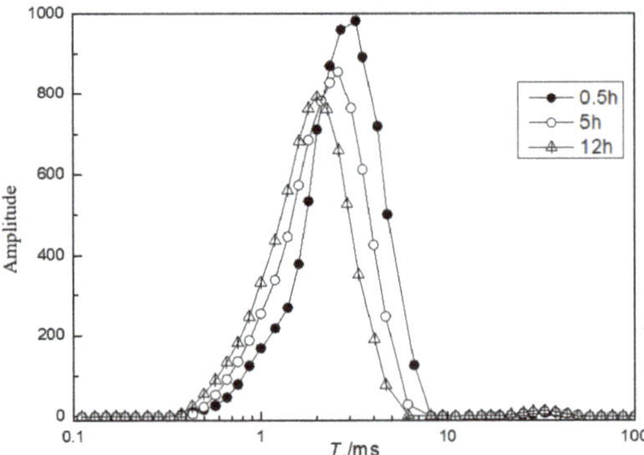

Figure 4. T_2 distribution of cement slurry with DMC content of 10%.

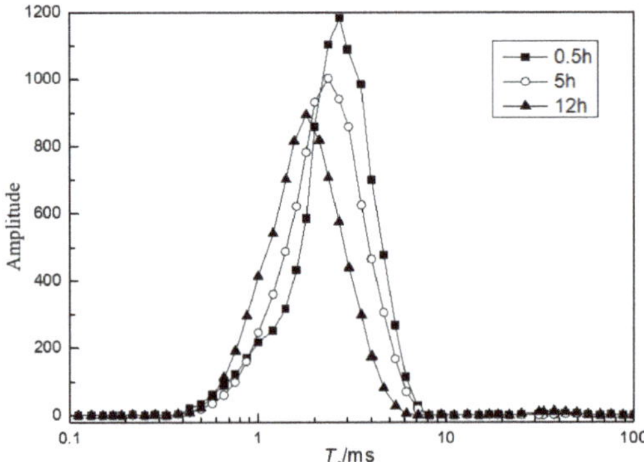

Figure 5. T_2 distribution of cement slurry with DMC content of 20%.

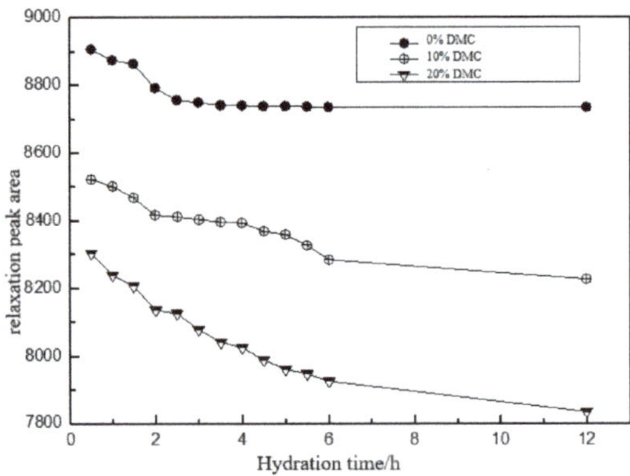

Figure 6. Changes in the relaxation peak area with time.

As shown in Figure 6, the T_2 peak area of cement slurry with retarders synthesized with different DMC dosages showed a relatively sharp decrease at the early stage of hydration (within the first 5 h), indicating that the cement slurry experienced a strong hydration reaction in the first 5 h. The trend then slowed down as the curing time increased. After 5 h, the peak area in the T_2 distribution of cement slurry containing retarder synthesized with 0% DMC essentially remained stable against the curing time. However, for cement slurry samples with retarders synthesized with 10% DMC and 20% DMC, the peak area continued to decrease after 5 h of curing time; the higher the DMC content, the greater the reduction in the peak area. The reduction in the peak area indicated that the water wrapped in the flocculant reacted with the flocculant and formed chemically bound water, promoting the extent of hydration reaction. The result shows that increasing the amount of cationic DMC in the retarder increased the relative percentage of the chemically bound water in the hydration products in the cement slurry. In other words, increasing the amount of DMC can promote hydration. This implies that DMC is conducive to improving the early strength of the cementing slurry in the low-temperature well section in the presence of a large temperature difference, preventing the excessively long retarding of the cement slurry in the low-temperature well section, and reducing the cementation quality of the cement slurry.

3.3. Effect of Cationic Dosage on Thickening Property and Compressive Strength of Cement Slurry

As shown in Figure 7, as the DMC dosage increased from 1% to 10%, the thickening time of cement slurry at 60 °C rose from 245 min to 350 min. The cement slurry was stable without settlement, and the compressive strength reached the maximum. With the further increase in the DMC dosage to 15% or 20%, the thickening time of the cement slurry decreased rapidly. Experimental observation indicates that there was significant stratification in the cement slurry at this DMC concentration, indicating that above the concentration of 10%, DMC will promote the excessive formation of early hydration products, resulting in instability, such as slurry flocculation and stratification and compromised retarding performance. Moreover, the contribution of early hydration products to the strength of the cement paste was not sufficient to offset the decline in cement paste strength caused by slurry settlement stratification, prohibiting the further improvement in strength.

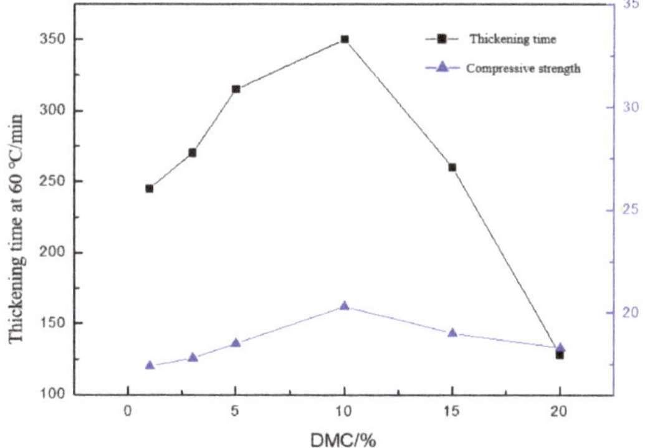

Figure 7. Effect of DMC dosage on the thickening time and compressive strength of cement slurry at 60 °C.

3.4. Effect of Temperature and Pressure on Thickening Performance of Cement Slurry

The thickening time of cement slurry was investigated by varying the experimental temperature and pressure to simulate the process where the temperature and pressure of the cement injection first increased and then decreased, as often encountered in oilfield application.

As illustrated in Figures 8 and 9, not only can polymeric ion retarders assure the thickening performance of cement slurry at the bottom of the well in the high-temperature and high-pressure well section, but they also adapt to the low-temperature and low-pressure section of the upper formation. At about 350 min and 480 min, the consistency of the cement slurry rose rapidly, which means that the right-angle thickening appeared on the distributions, and the slurry had a great thickening property. This can be attributed to DMC contribution to the formation of the early hydrate, which successfully ensured the cementing quality of low-temperature well sections.

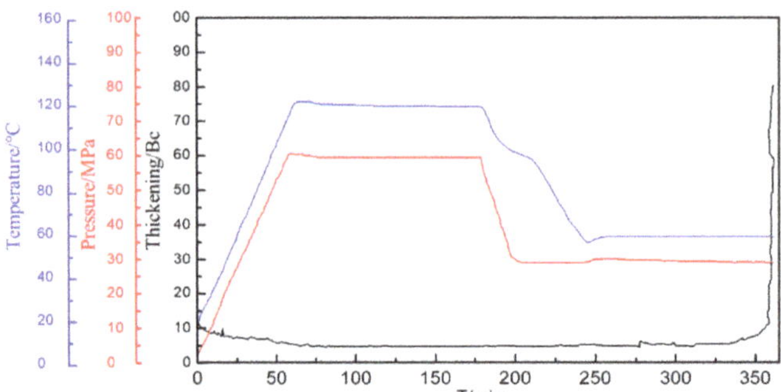

Figure 8. Thickening curve of polymeric ion retarder from 120 °C to 60 °C.

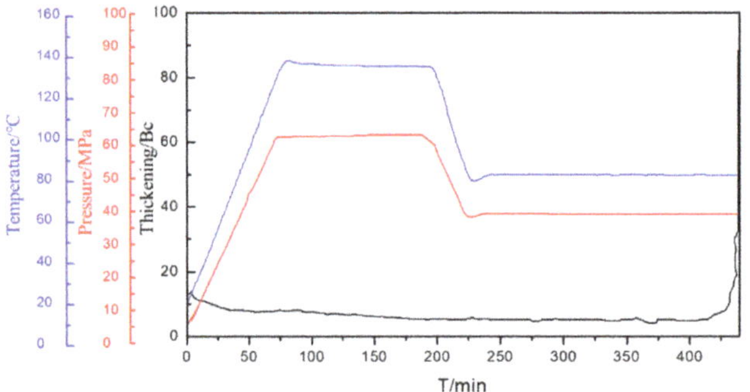

Figure 9. Thickening curve of polymeric ion retarder from 150 °C to 90 °C.

The modified retarder in this paper is mainly used to narrow the gap between the setting time of the cement slurry in the top section of the well with low temperature and the bottom section of the well with high temperature. If the conventional retarder is added, the cement slurry in the low temperature environment will not solidify for 24 h, while the cement slurry in the high temperature environment may solidify in 3–5 h. It can be seen from Figures 8 and 9 that the cement slurry with the polymer ion retarder is simulated to return to the low temperature environment at the upper part of the well from the bottom of the well with high temperature. The thickening time of the cement slurry is 6–7 h, which is equivalent to the thickening time under the high-temperature environment at the bottom of the well. It ensures that the cement slurry of the whole well section maintains a similar hydration rate, while improving the cementing quality.

3.5. XRD Analysis of the Influence of Polymeric Ion Retarder on the Hydration Performance of Cement

The four kinds of cement slurry without retarder and with retarder containing 0% DMC, 10% DMC, and 20% DMC were cured at 90 °C for 24 h, and then the X-ray diffraction analyzer was used to analyze the components of the cement. The results are shown in Figure 10.

As shown in Figure 10, in blank samples and samples containing retarders, the characteristic diffraction peaks of Ca $(OH)_2$, C-S-H, and ettringite (AFT) can be identified in all the XRD patterns, indicating that retarders did not affect the composition of cement hydration products. However, the diffraction peak intensity of the blank sample was the highest in four patterns, followed by the sample with 20% DMC in the retarder. With the decrease in DMC content, the intensity of the characteristic diffraction peaks of $Ca(OH)_2$, C-S-H, and AFT gradually decreased, indicating that the retarder inhibited the formation of a hydration product. What this also indicates is that the introduction of DMC into the retarder could facilitate the formation of $Ca(OH)_2$ crystals and C-S-H gel, promoting the early strength of cement slurry, making it possible to meet the requirement of the strength of the cement slurry during the curing time in low-temperature well sections under the condition of a large temperature difference.

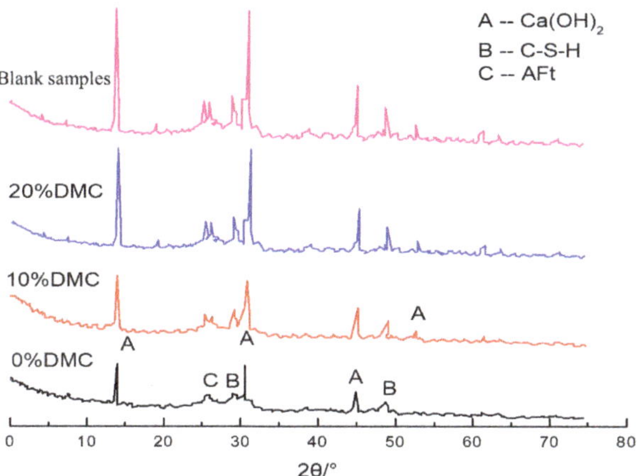

Figure 10. XRD patterns of cured cement with different retarders.

4. Conclusions

1. The water phases in the early hydration stage of cement slurry, wrapped water in precipitate and free water, were characterized using NMR, which shows that the former is the major reaction phase.
2. Compared to the blank cement slurry, the relaxation time T_2 of the cement slurry with a retarder shifted to the right, and the relaxation duration, relaxation peak value, and relaxation area dropped during the hydration process. This indicates that the retarder reduced the van der Waals and electrostatic forces as well as the hydration reaction rate between the wrapped water in the precipitate and the cement particles, resulting in low early cement strength and a longer hardening time.
3. The addition of cationic DMC to the retarder promoted the amount of chemically bound water in the hydration products of the cement slurry, thereby improving the cement slurry's early hydration, reducing the prolonged curing of the cement slurry containing a conventional retarder at lower temperatures, and improving the early strength of the cementing in the upper low-temperature well section with a large temperature difference.
4. XRD analysis results show that introducing cationic DMC into the retarder could promote the formation of $Ca(OH)_2$ crystal and C-S-H gel, enhancing the early strength development of the cement slurry, and ensuring the cement slurry strength in low-temperature well sections.

Author Contributions: Formal analysis, Y.C.; Resources, H.Y.; Data curation, H.G.; Writing—original draft, Z.Q.; Writing—review & editing, C.H. and Q.Y. All authors have read and agreed to the published version of the manuscript.

Funding: Dense reservoirs long-term development of the cementing technology research.

Data Availability Statement: The authors confirm that the data supporting the findings of this study are available within the article.

Conflicts of Interest: We declare that we have no financial and personal relationships with other people or organizations that can inappropriately influence our work, there is no professional or other personal interest of any nature or kind in any product, service and/or company that could be construed as influencing the position presented in, or the review of, the manuscript entitled.

References

1. Su, R.; Li, Q. Study and application of high temperature retarder GH-9 for cement slurry. *Drill. Fluid Complet.* **2005**, *22*, 89–92.
2. Rigby, T.L. Gas Migration in Oilwell Cements. Ph.D. Thesis, Howard University, Washington, DC, USA, 2008.
3. Liu, K.; Cheng, X.; Zhang, X.; Guo, X.; Zhuang, J. Design of low-density cement optimized by cellulose-based fibre for oil and natural gas wells. *Powder Technol.* **2018**, *338*, 506–518. [CrossRef]
4. Guo, S. Key Technology Research of Slurry System Suited for Long Cementing Interval in Deep Well. Ph.D. Thesis, China University of Petroleum (East China), Qingdao, China, 2013.
5. Zhao, C.; Zhang, C.; Sun, H. Development and application of the well cement retarder SD210L with a wide high temperature range. *Nat. Gas Ind.* **2013**, *33*, 90–94.
6. Guo, Z.; Liu, Z.; Liao, K.; Yang, X. Research status of synthetic oil well cement retarder. *Recent Pat. Eng.* **2021**, *15*, 53–59.
7. Zhang, L.; Zhuang, J.; Liu, H.; Li, H.M.; Zhao, Z. Terpolymerization and performance of 2-acrylamide-2-methyl propane sulfonic acid/itaconic acid/N-vinyl-2-pyrrolidone. *J. Appl. Polym.* **2010**, *117*, 2951–2957. [CrossRef]
8. Zhao, H.; Ye, T.; Wang, Q. Laboratory synthesis and evaluation of a new high-temperature retarder PC-H42L. *J. Yangtze Univ. (Nat. Sci. Ed.)* **2012**, *9*, 73–74.
9. Guo, S.; Bu, Y.; Shen, Z.; Du, F.; Ma, C. The Effect of Synthesis conditions on the performance of large temperature difference retarder. *Res. J. Appl. Sci. Eng. Technol.* **2013**, *5*, 4751–4756.
10. Guo, J.; Xia, X.; Liu, S.; Jin, J.; Yu, Y.; Yu, Q. A high temperature retarder HTR-300L applied in long cementing interval. *Pet. Explor. Dev.* **2013**, *40*, 656–660. [CrossRef]
11. Qin, W. Cementing technological present situation and development direction. *Oil Forum* **2007**, *2007*, 6–11.
12. Li, S.; Tu, R. Analysis of long interval isolation cementing technology. *Guide Bus.* **2012**, *2012*, 276–277.
13. Zhao, B.; Zou, J.; Liu, A. Research and application on new type of retarder used in long cementing intervals. *Drill. Fluid Complet. Fluid* **2011**, *28*, 10–12.
14. Al-Yami, A.S.; Al-Arfaj, M.K.; Nasr-El-Din, H.A. Development of new retarder systems to mitigate differential cement setting in long deep liners. In Proceedings of the SPE/IADC Middle East Drilling and Technology Conference, Cairo, Egypt, 22–24 October 2007; p. 107538.
15. Zhou, B.; Han, Q.; Yang, P. Characterization of nanoporous systems in gas shales by low field NMR cryoporometry. *Energy Fuels* **2016**, *30*, 9122–9131. [CrossRef]
16. Yan, J.P.; He, X.; Geng, B.; Hu, Q.H.; Feng, C.Z.; Kou, X.P.; Li, X.W. Nuclear magnetic resonance T_2 spectrum: Multifractal characteristics and pore structure evaluation. *Appl. Geophys.* **2017**, *14*, 205–215. [CrossRef]
17. Adams, A. Analysis of solid technical polymers by compact NMR. *Trends Anal. Chem.* **2016**, *83*, 107–119. [CrossRef]
18. Blümich, B.; Singh, K. Desktop NMR and its applications from materials science to organic chemistry. *Angew. Chem.* **2018**, *57*, 6996–7010. [CrossRef] [PubMed]
19. Blümich, B. Introduction to compact NMR: A review of methods. *TrAC Trends Anal. Chem.* **2016**, *83 Pt A*, 2–11. [CrossRef]
20. Blümich, B.; Perlo, J.; Casanova, F. Mobile single-sided NMR. *Prog. Nucl. Magn. Reson. Spectrosc.* **2008**, *52*, 197–269. [CrossRef]
21. She, A.-m.; Yao, W. Research on hydration of cement at early age by proton NMR. *J. Build. Mater.* **2010**, *13*, 376–379.
22. Zhu, D.; Deng, Z.; Chen, S. A review of nuclear magnetic resonance(NMR) technology applied in the characterization of polymer gels for petroleum reservoir conformance control. *Pet. Sci.* **2021**, *18*, 1760–1775. [CrossRef]
23. Guo, J.C.; Zhou, H.Y.; Zeng, J.; Wang, K.J.; Lai, J.; Liu, Y.X. Advances in low-field nuclear magnetic resonance(NMR) technologies applied for characterization of pore space inside rocks: A critical review. *Pet. Sci.* **2020**, *17*, 1281–1297. [CrossRef]
24. Xiao, J.-m.; Fan, H.-h. Application of solid-state nuclear magnetic resonance techniques in research of cement and hydration products. *J. Mater. Sci. Eng.* **2016**, *34*, 166–172. [CrossRef]
25. She, A.-m.; Yao, W. Characterization of microstructure and specific surface area of pores in cement paste by low field nuclear magnetic resonance technique. *J. Wuhan Univ. Technol.* **2013**, *28*, 11–15.
26. Sharma, S.; Casanova, F.; Wache, W.; Segre, A.; Blümich, B. Analysis of historical porous building materials by the NMR-MOUSE. *Magn. Reson. Imaging* **2003**, *21*, 249–255. [CrossRef] [PubMed]
27. Strange, J.H.; Webber, J.B. Characterization of porous solids by NMR. *Magn. Reson. Imaging* **1995**, *12*, 257–259. [CrossRef]
28. Cheng, F.Z.; Lei, X.W.; Meng, Q.S.; Chen, J. Pore Water content of solidified dredging silt and its distribution based on nuclear magnetic resonance. *J. Yangtze River Sci. Res. Inst.* **2016**, *33*, 116–120.
29. Xiao, J.-M.; Fan, H.; Wu, Y.; Zeng, M. 29Si Solid-state high resolution NMR analysis of cement clinkers calcined by sewage sludge ash instead of clay. *J. Mater. Sci. Eng.* **2016**, *34*, 460–464.
30. Ran, Q.-P.; Liu, J.-P.; Zhou, D.; Mao, Y.; Yang, Y. Research on effects and mechanism of amphoteric comb-like copolymers on dispersion and its retention of cement pasters. *J. Build. Mater.* **2012**, *15*, 158–162.
31. Miljkovic, L.; Lasic, D.; MacTavish, J.C.; Pintar, M.M.; Blinc, R.; Lahajnar, G. NMR studies of hydrating cement: A spin–spin relaxation study of the early hydration stage. *Cem. Concr. Res.* **1988**, *18*, 951–956. [CrossRef]
32. Bohris, A.J.; Goerke, U.; McDonald, P.J.; Mulheron, M.; Newling, B.; Le Page, B. A broad line NMR and MRI study of water and water transport in Portland cement pastes. *Magn. Reson. Imaging* **1998**, *16*, 455–461. [CrossRef]
33. Zhang, R.; Wang, J.; Peng, Z. Effects of retarder on evolution process of water in cement paste. *Fine Chem.* **2016**, *33*, 1416–1421.

MDPI

Article

Experimental Study of Proppant Placement Characteristics in Curving Fractures

Zhiying Wu, Chunfang Wu and Linbo Zhou *

State Key Laboratory of Shale Oil and Gas Enrichment Mechanisms and Effective Development, Beijing 100101, China

* Correspondence: xiangao729@163.com

Abstract: Proppant placement in hydraulic fractures is crucial for avoiding fracture closure and maintaining a high conductivity pathway for oil and gas flow from the reservoir. The curving fracture is the primary fracture form in formation and affects proppant–fluid flow. This work experimentally examines proppant transport and placement in narrow curving channels. Four dimensionless numbers, including the bending angle, distance ratio, Reynolds number, and Shields number, are used to analyze particle placement in curving fractures. The results indicate that non-uniform proppant placement occurs in curving fractures due to the flow direction change and induces an irregular proppant dune. The dune height and covered area are lower than that in the straight fracture. The curving pathway hinders proppant distribution and leads to a dune closer to the inlet. When the distance increases between the inlet and curving section, a large depleted zone in the curving section will be formed and hinder oil and gas flowback. The covered area has negative linear correlations with the Reynolds number and Shields numbers. Four dimensionless parameters are used to develop a model to quantitatively predict the covered area of particle dune in curving fractures.

Keywords: hydraulic fracturing; proppant dune; proppant placement; complicated fracture; multiphase flow

Citation: Wu, Z.; Wu, C.; Zhou, L. Experimental Study of Proppant Placement Characteristics in Curving Fractures. *Energies* **2022**, *15*, 7169. https://doi.org/10.3390/en15197169

Academic Editor: Daoyi Zhu

Received: 27 July 2022
Accepted: 16 September 2022
Published: 29 September 2022

1. Introduction

Hydraulic fracturing has become an essential method to create complex fractures in unconventional reservoirs to improve oil and gas recovery. Proppants should be transported by fracturing fluid into the fracture [1,2]. Sufficient proppants can avoid fracture closure and maintain a high conductivity pathway for liquid flowback from the reservoir. Many factors, including the fracture shape, injection parameters, and proppant properties, change particle placement [3–5]. Experimental results have proven that pre-existing cracks significantly affect hydraulic fracture propagation. The hydraulic fracture probably deviates into a pre-existing fracture during the intersection and then turns to the original direction. Typically, a hydraulic fracture is curved and irregular [6–10], and the complicated pathway probably affects particle–fluid flow, which changes particle distribution [11–13]. A specific understanding of proppant placement characteristics in a curving fracture with bends is crucial to optimizing fracturing design and enhancing the stimulation effect for unconventional reservoirs.

Various narrow channels were developed based on the similarity theory of fluid mechanics to study proppant placement in hydraulic fracture. For simplicity, vertical planar channels with narrow widths have been used to represent hydraulic fractures for decades. A straight channel containing two parallel acrylic panels was first set up by Kern et al. [14]. Sands and water were injected into the channel. It was observed that particle suspension and settling were two primary transport mechanisms in the straight fracture. An initial sand dune would form near the inlet due to particle settling. The injected sands flowed through the initial dune and deposited to the backside for the dune development toward the fracture tip. The dune would be in equilibrium, and the subsequently injected sands were suspended deeper into the experimental channel. Babcock et al. [15] used a planar channel to study the sand buildup mechanisms. Tap water and three types of sand were employed in the experiment. It was found that the equilibrium bed height would significantly vary as the fluid velocity increases or decreases. A correlation equation was developed for the prediction of the equilibrium height and velocity related to the proppant type and size, the fluid properties, and the fracture geometry. According to the experimental channel at the STIM-LAB facilities, Wang et al. [16] developed a bi-power law model to calculate the equilibrium height. Water and low viscous fluid were used as fracturing fluids. The equation was widely used to model particle transport in vertical fracture and benchmark numerical simulation models [17–20]. Since particle settling would be reduced in the fracture with a small aperture, Zeng et al. [21] studied the particle dune shape in a super-narrow channel and developed two correlations to predict the dune height and coverage area. It was found that the dune in narrow fracture was higher than the wider one. Qu et al. [4] studied proppant placement in a wedge fracture. Low viscous fluid with 1.5 mPa·s was prepared as fracture fluid. The experimental channel with a contracted aperture was achieved by gradually inserting acrylic panels to gradually reduce the channel width. A regression model with four dimensionless parameters was proposed to predict the dune equilibrium height. Recently, other fracturing fluids, such as high viscosity guar gum, foamed fluid, and supercritical carbon dioxide, were tested in the straight channel to study proppant placement [5,22,23].

Because hydraulic fracturing can easily cause a complex fracture in the reservoir-developed fissures, complex experimental channels were developed to understand proppant placement in complicated fracture systems. Two types of channel systems have been proposed so far. A planar channel has several secondary and tertiary channels [13,24,25]. Sahai et al. [25] first developed a complex channel that contained a primary channel, three secondary channels, and two tertiary channels. The location of the subsidiary channels from the inlet impacted the proppant-covered area. It was found that the closest subsidiary fractures to the inlet received more proppant than the farthest fractures. However, the propped area in the tertiary fracture slots was considerably less than in the primary and secondary ones. Two transport behaviors are found as the proppant flows into the secondary channel. Proppants roll down to the secondary channel by the gravity effect. Furthermore, Proppants were suspended in the secondary channel by vortices as the fluid velocity was high enough. Another channel is that straight channels have one or two-branched channels [19,26]. Particles mainly settled in the primary channel to form a high dune and were more likely to accumulate in the branched channel near the inlet. The turbulent flow and eddies at fracture intersections could decrease the particle dune height and improve particle transport deeper into narrow-branched fractures. [11,19,27–29].

Overall, previous literature studied proppant–fluid flow in planar fractures and complex fracture networks. Typically, curving fractures are the most common fracture in the reservoir due to the influence of formation heterogeneity and pre-existing fractures. Fracture shapes are more likely to be nonlinear. However, proppant placement characteristics in a curving fracture are still not well understood. Furthermore, there is no equation to quantitatively predict the coverage area of the proppant dune. For this reason, the present article's objective is to study the effect of curving fractures on proppant placement, particle size distribution, and particle dune area. Seven narrow channels with a curving section are set up to analyze proppant placement. A laser particle analyzer measures the particle size distributions in different positions of channels. Four dimensionless parameters are used to develop the relationship with the dune-covered area. A regression model is developed to calculate the percentage of covered area in curving fractures quantitatively. It is beneficial to understand proppant placement and fracturing design.

2. Experimental System and Method

2.1. Experimental Method

Figure 1 shows the experimental system for proppant transport and placement in a curving channel. Natural sands are typically used as proppant in hydraulic fracturing [30], and four types of sands are used in the experiment, including 10/20 mesh, 20/40 mesh, 40/70 mesh, and 70/140 mesh. Table 1 lists the sand size and properties. Sands are mixed with tap water in the tank, and the sand flow rate is controlled by a funnel. A slurry pump with a variable frequency driver (VFD) is used to meet the requirements of the flow rates. To ensure accuracy, we calibrated the rate by collecting water at the outlet of the channel for a constant interval, then adjusted the rate by the VFD. During the experiment, the slurry is pumped into the channel by five holes on the channel inlet. Three cameras with 5472 × 3648 resolution record particle distribution. A 160-mesh sieve collected sands that flowed out of the channel at the outlet during the experiment. When the dune is equilibrium, the test is terminated. Once the test is complete and the channel is opened. Sand samples are collected from each individual region. Particle size distribution is measured by a laser particle analyzer.

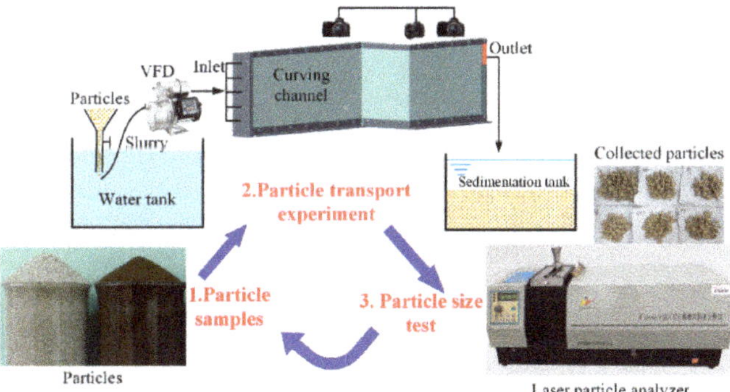

Figure 1. Experimental procedures and instruments.

Table 1. Particle parameters used in experiments.

Particle	Mesh	Size Range (mm)	Average Diameter d50 (mm)	Apparent Density ρp (kg/m^3)	Repose Angle θr (°)
Sand	10/20	2/0.84	1.41	2650	34.8
	20/40	0.84/0.42	0.633	2650	33.5
	40/70	0.42/0.212	0.311	2650	37.3
	70/140	0.212/0.053	0.155	2650	40.2

Figure 2 shows the structure and dimensions of a curving channel. To ensure the accuracy of the bending angle and the sealing at the bend, two whole acrylic sheets are bent twice at the required angle. A blue sticker is attached to one side of the channel as the background. The experimental channel is divided into three parts: the first straight section, the curving section, and the second straight section. The total length, height, and width are 1.5 m, 0.27 m, and 4 mm. Five injection holes are on the inlet boundary of the channel to simulate the perforation cluster; the hole diameter is 9 mm. The outlet is at the top right corner, which is similar to the channels [25,31]. The outlet height is 70 mm. The seven channels have the same inlet and outlet configuration to use the pumping system.

In Figure 2a, the bending angle varies from 45° to 180°, and other dimensions are kept constant. Four curving channels with varied bending angles are assembled. In Figure 2b, the bending angle is 90°, and the length of the first straight section varies from 0.25 m to 1.0 m. Correspondingly, the length of the second straight section changes from 1.0 m to 0.25 m. Four curving channels with varied distance ratios are used in the experiments.

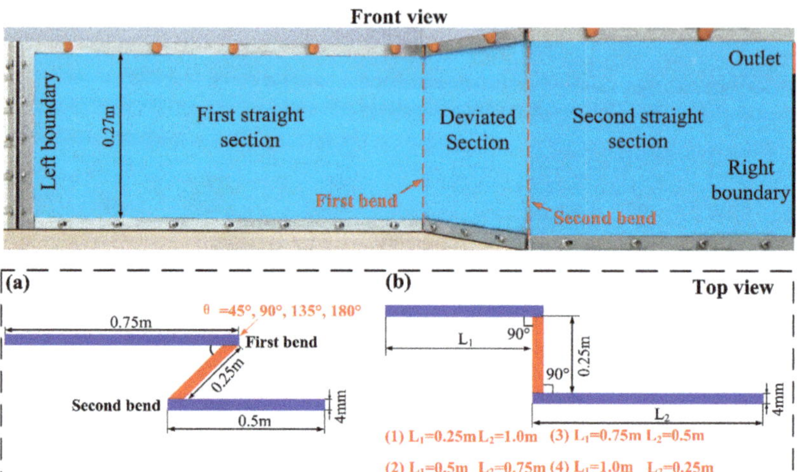

Figure 2. Curving channels at the front view and configurations at the top view.

Since the inner sizes of all channels are the same, 24 region codes from A1 to F4 are marked for describing the particle size and distribution, as shown in Figure 3.

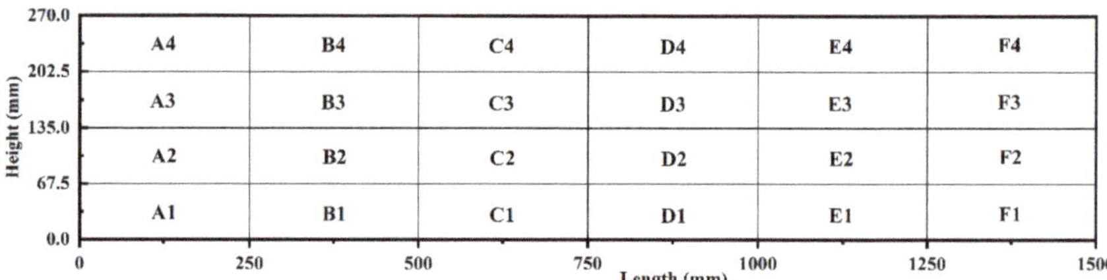

Figure 3. The region codes of the channel for particle size analysis.

2.2. Sands and Fluid

Figure 4 shows four sizes of natural sands used in the present study, including 10/20 mesh, 20/40 mesh, 40/70 mesh, and 70/140 mesh. The size range is measured by the laser particle analyzer. The median-diameter d50 is a commonly used parameter to represent the average particle size [32]. It is defined as the particle size at which the cumulative volume percentage of particles reaches 50%. Table 1 lists the relevant physical properties of sands.

Figure 4. Particle samples used in experiments.

The sand mixes with tap water to prepare the slurry. The average slurry velocity in a fracture without proppant deposition typically ranges from 0.1 m/s to 0.5 m/s [25], the experimental velocity is from 0.12 m/s to 0.5 m/s.

2.3. Experimental Parameters

The percentage of the dune coverage area (*PCA*) can be used to evaluate particle placement, expressed by Equation (1).

$$PCA = \frac{PDA}{FCA} \times 100\% \quad (1)$$

where the *PDA* is the proppant dune area perpendicular to the width direction; *FCA* is the fracture cross-section area perpendicular to the width direction. The dimensionless numbers are typically used to set up the correlation for the fluid experiment [33]. The *PCA* has a relationship with the bending angle, the distance between the bending section and inlet, fluid velocity, and particle size. Thus, the dimensionless expression is as follows.

$$PCA = f\left(\theta, \frac{L_1}{L}, Re, S\right) \quad (2)$$

where θ is the bending angle; L_1 is the length of the first straight section; L is the total length of the channel; The Reynolds number of *Re* is expressed in Equation (3); The Shields number of S is expressed by Equation (5).

The Reynolds number *Re* is usually used to represent flow regimes, including turbulence and laminar flow.

$$Re = \frac{2\rho_f v_i l_h}{\mu_f} \quad (3)$$

where ρ_f is the fluid density, kg/m^3; μ_f is the fluid viscosity, Pa·s; v_i is the fluid velocity, m/s; l_h is the hydraulic diameter of the narrow channel.

$$l_h = \frac{2wh}{w+h} \quad (4)$$

where w is the channel width, m; h is the channel height, m.

Shields number represents the shear stress ratio at the dune top to the particle weight to analyze particle motion. At very low Shields Number, any settled proppant remains stationary and is hardly transported by fluid. Mack et al. [34] proposed the Shields number equation to represent particle transport in a narrow rectangular channel, expressed in Equation (5).

$$s = \frac{8v_i\mu_f}{(\rho_p - \rho_f)gd_p w} \quad (5)$$

where d_p is the mean particle diameter, m; ρ_g is particle density, kg/m^3; g is the gravitational acceleration, m/s^2.

Table 2 list the experimental parameters and dimensionless numbers. The *Re* is from 946 to 3942, and the S is from 0.04 to 0.32. Both parameters are in the typical range [35,36].

Table 2. Experimental parameters and dimensionless numbers.

Case	Fluid Phase	Particle Phase	Dimensionless Number			
	v_i (m/s)	d_{50} (mm)	L_1/L	θ (°)	Re	S
1	0.42	0.311	0.5	45/90/135/180	3311	0.16
2	0.42	0.311	0.16/0.33/0.5/0.66	90	3311	0.16
3	0.12	0.311	0.5	45/90/135/180	946	0.05
4	0.12	0.311	0.16/0.33/0.5/0.66	90	946	0.05

Table 2. Cont.

Case	Fluid Phase	Particle Phase	Dimensionless Number			
	v_i (m/s)	d_{50} (mm)	L_1/L	θ (°)	Re	S
5	0.21	0.311	0.5	45/90/135/180	1655	0.08
6	0.21	0.311	0.16/0.33/0.5/0.66	90	1655	0.08
7	0.5	0.311	0.5	45/90/135/180	3942	0.2
8	0.5	0.311	0.16/0.33/0.5/0.66	90	3942	0.2
9	0.42	0.155	0.5	45/90/135/180	3311	0.32
10	0.42	0.155	0.16/0.33/0.5/0.66	90	3311	0.32
11	0.42	0.633	0.5	45/90/135/180	3311	0.08
12	0.42	0.633	0.16/0.33/0.5/0.66	90	3311	0.08
13	0.42	1.41	0.5	45/90/135/180	3311	0.04
14	0.42	1.41	0.16/0.33/0.5/0.66	90	3311	0.04

3. Results and Analysis

3.1. Effect of the Angle, θ

Figure 5 shows four 40/70 mesh sand dunes at equilibrium for Case 1 (Table 2). For convenient comparison, three sections of the curving channel are placed in the plane. Two red dash lines in the figure mark the boundary of the curving section. The first straight section and the second straight section are located on the curving section's left and right sides. It should be noted that brown particles as a tracer agent are injected twice during the experiments for demarcating the dune into three regions, representing three injection sequences of S_1, S_2, and S_3. Figure 5a shows a sand dune in the straight channel. The region of S_1 shows the initial dune shape formed by the first injection sequence. The later injected sands are lifted to overshoot the front region of S1 and settle to the backside, forming the regions of S_2 and S_3 in sequence, which are identical to experimental results [14,25,37,38]. As the sand dune is in equilibrium, the dune height and shape are constant.

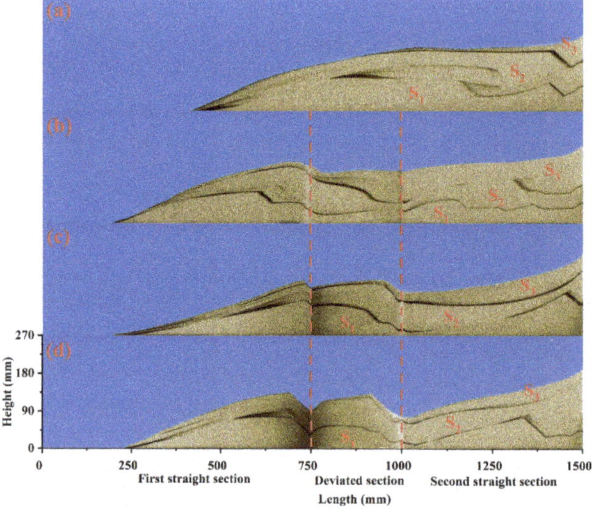

Figure 5. Snapshots of 40/70 mesh sand dunes in four channels at the stable status at $L_1/L = 0.5$, *Re* = 3311, and S = 0.16, front view: (**a**) The straight channel, θ = 180°; (**b**) The curving channel, θ = 135°; (**c**) The curving channel, θ = 90°; (**d**) The curving channel, θ = 45°.

The curving pathway significantly affects particle transport behaviors compared to the straight channel. More particles deposit in the first straight section and form a dune closer to the left boundary. It indicates that the curving section can hinder particle transport. The dune buildup in the curving section originates from two issues, as the following: (1) particle rolling from the first straight section; (2) particle settling in the suspension after colliding with the curving section. Although hindered by the curving section, many particles could be suspended through two bends and deposited along with the second straight section. Meanwhile, some particles directly flow out of the channel during the experiment. Another important finding is that the particle resuspension appears around two bends as the dune increase to a threshold value close to the equilibrium height, as shown in Figure 6. According to the observation, there are different scale vortices around the bends, and they are kept in the clockwise direction due to the turbulent stress [39]. The vortices can resuspend the settled particles through the curving section. Due to the vorticity around two bends, there are two depressions on the top of the dune.

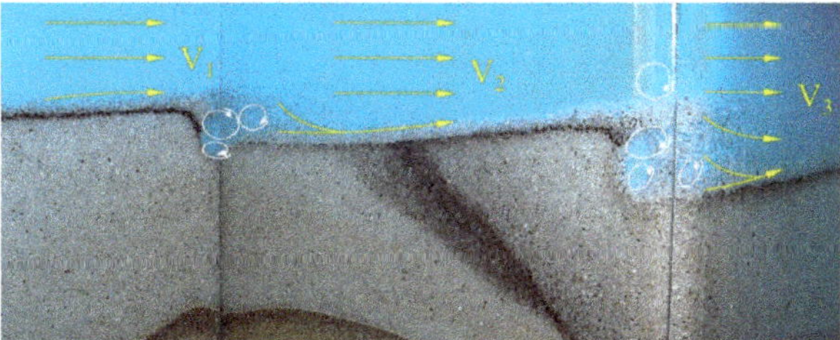

Figure 6. Vortices of 40/70 mesh sands around two bends, enlarged view of the deviated section in Figure 5c.

Once the test is complete and the channel is opened. Sand samples are collected from each individual region, illustrated in Figure 3. Particle size distribution is measured by a laser particle analyzer. Then, the contour map of the size distribution is obtained by the software of MATLAB 9.0, as shown in Figure 7. Particle sizes gradually reduce from the inlet to the outlet, and large particles accumulate in the front of the sand dune. In the straight channel, 65/70 mesh sands are in the E1 and F1 regions, and more than 70 mesh sands are in the F1. The result means small particles are easily transported deeper into the fracture during the first sequence of S_1. In the second last row from C2 to F2, they represent the second injection sequence of S_2, extensive sands sized from 40 to 60 mesh settle these regions. Furthermore, the particle size in the regions from D3 to F3 ranges from 40 to 50 mesh. Large sands deposit on the top of the dune at the late injection sequence, and smaller particles are out of the channel. The reason is that the decrease in the flow gap increases the fluid velocity, transporting larger particles out of the channel. As the bending angle decreases, it is founded that the mean particle sizes in E1 and F1 increase. For the angle of 45°, sand sizes range from 50 to 62 mesh in E1 and F1. It is interesting to note that there are more large sands in the straight section due to the curving section's hindrance. The particle size mainly ranges from 50 to 55 mesh in the curving section.

Four sand samples are collected at the outlet by a 160-mesh sieve when the injection time is 180 s. Figure 8a shows the particle size distributions of four collected samples. When the angle is 135°, the mean particle diameter of d_{50} is 303 μm, and the diameter increases to 317 μm at $\theta = 45°$, which is larger than the original mean diameter of 311 μm. For the straight channel, smaller sands ($d_{50} = 282$ μm) can flow out of the channel, and large particles deposit in the channel for the dune buildup, consistent with the experimental results [25,40]. In Figure 8b, it is found that the PCA decreases with the reduction of the bending angle. The PCA has a nonlinear correlation with the dimensionless number of angles. Fitting a natural logarithmic law trend through four data obtains a coefficient of determination, R^2 of 0.89.

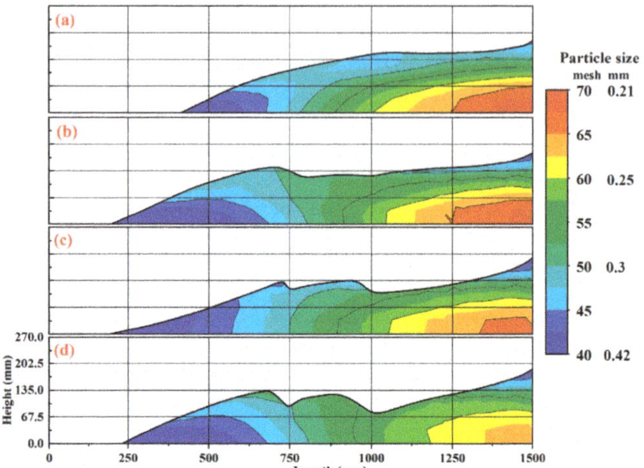

Figure 7. Mean particle size distribution across four channels: (**a**) The straight channel, θ = 180°; (**b**) The curving channel, θ = 135°; (**c**) The curving channel, θ = 90°; (**d**) The curving channel, θ = 45°.

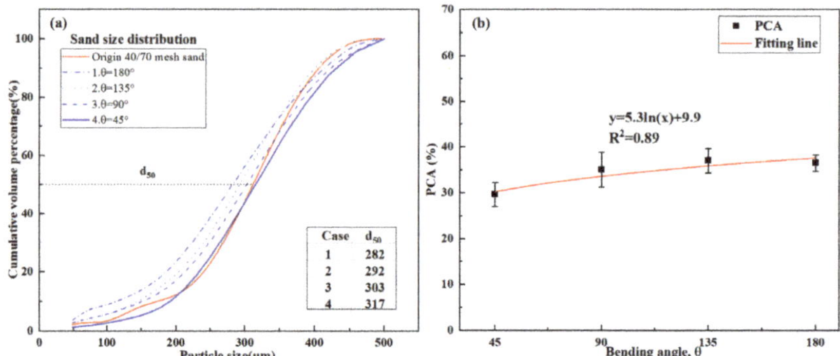

Figure 8. Effect of the bending angle on the particle dune distribution: (**a**) The particle size distributions of sands flowed out the channel at the injection time = 180 s; (**b**) The PCA in four channels with different bending angles.

In conclusion, as the bending angle decreases, a curving pathway induces fluid flow redirection and turbulent flow, enhancing particle transport capacity. As a result, more large particles are washed out of the channel. Moreover, the dune shape is more irregular and smaller.

3.2. Effect of the Distance Ratio, L_1/L

Since the location of the natural fracture is random, the distance between the natural fracture and the wellbore would be variable and affects the particle–fluid flow in the curving fracture. Figure 9 shows the 40/70 mesh sand dunes in four channels for Case 2 (Table 2). Two red dash lines in the figure mark the boundary of the curving section. It should be noted that brown particles as a tracer agent are injected twice during the experiments for demarcating the dune into three regions, representing three injection sequences of S_1, S_2, and S_3. In Figure 9a,b, there is a large depleted zone without sands between the left boundary and the front side of the particle dune. In Figure 9a, almost all sands are transported through the curving section and deposited in the second straight section. However, the fracture will close in the depleted zone because there are no proppants to prop the fracture, and the closure region will significantly increase flow friction during production. According to the size of the depleted zone, it is found that the curving section has a hindering effect on particle transport as the ratio increases. As the ratio is more than 0.33, the dune's front side is closer to the inlet, and more sands deposit in the first straight section.

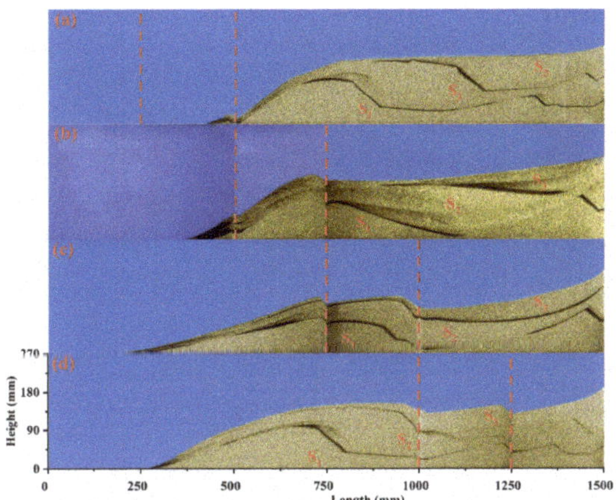

Figure 9. Snapshots of 40/70 mesh sand dunes in four channels at equilibrium at θ = 90°, *Re* = 3311 and S = 0.16, front view: (**a**) $L_1/L = 0.16$; (**b**) $L_1/L = 0.33$; (**c**) $L_1/L = 0.5$; (**d**) $L_1/L = 0.66$.

Figure 10 shows the contour map of the particle size distribution in four channels with different distance ratios. In Figure 10a, the particle size distribution in the second straight section is similar to that in the straight channel. After the injected particle flows through the curving section, large particles settle in the front, and the small particles deposit in the end. For the late injected sequences of S_2 and S_3, more large particles build up the dune in the channel. As the distance ratio increases, the curving section is far away from the entrance, and more large particles sized from 40 to 50 mesh are hindered in the first straight section. Furthermore, the smaller particles hardly deposit at the end of the channel. For example, Figure 10d shows that the sand size in the F1 is from 60 to 65 mesh. The probable explanation is that the vortex flow around two bends is closer to the outlet, so smaller particles are prone to flow out of the channel by the vortex.

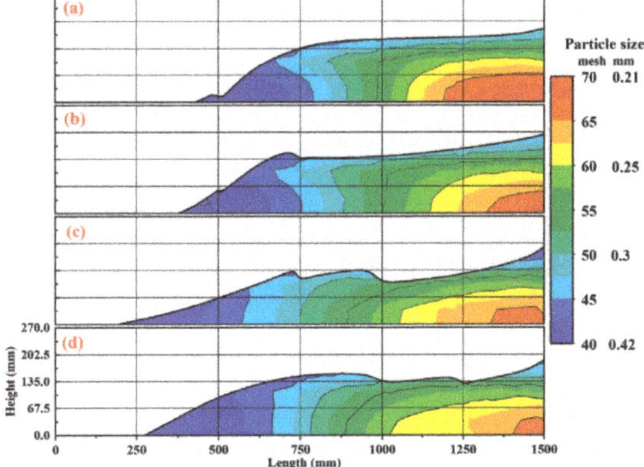

Figure 10. Mean particle size distribution across four channels: (**a**) $L_1/L = 0.16$; (**b**) $L_1/L = 0.33$; (**c**) $L_1/L = 0.5$; (**d**) $L_1/L = 0.66$.

Figure 11a shows the particle size distributions for the sand deposited in the sedimentation tank when the injection time is 180 s. Compared to the original distribution, the bigger particle could flow out of the channel with a smaller distance ratio. When the ratio is 0.166, the mean diameter is 316 μm. When the ratio increases to 0.66, the diameter decreases to 291 μm. Figure 11b shows a natural logarithmic law trend between the PCA and distance ratio.

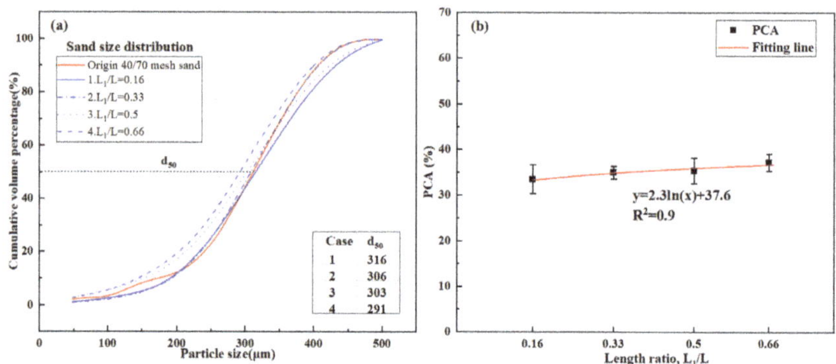

Figure 11. Effect of the distance ratio on the particle dune distribution: (**a**) The particle size distributions of sands flowed out the channel at the injection time = 180 s; (**b**) The PCA in four channels with different distance ratios.

3.3. *Effect of the Reynolds Number, Re*

Since the particle transport originated from the fluid flow, the flow pattern would affect the particle distribution. Figure 12 shows the front view of 40/70 mesh sand distributions in four channels for Case 3 (Table 2), the slurry flow shown in Figure 12 is laminar. Two red dash lines in the figure mark the boundary of the curving section. The 40/70 mesh sand builds up a higher dune in the straight channel, closer to the left boundary. For the curving channel, three dunes contact the entrance. In Figure 12c,d, the lowest injection hole was totally blocked during the experiment. The curving pathway has a more hindering effect on particle deposition in laminar flow with a low Reynolds number, especially in the channel with small bending angles. The flow would change to turbulence when the dune increases to a threshold height, and then the vortex flow appears around two bends. Due to the vortex wash, depression will be formed around the bend. The smaller the bending angle, the larger the depression.

Figure 13 shows the front view of 40/70 mesh sand distributions in four channels for Case 4 (Table 2). It is found that the distance ratio has little effect on particle distribution when the flow is laminar. Particle settlement and bedload dominate the transport mechanisms [37]. As a result, the injected particles quickly settle near the entrance and accumulate a high dune in the channel.

In Figure 14, Re has a negative linear correlation with PCA. The experiments are Cases 1 to 8. The higher the Reynolds number, the lower the covered areas, especially in the curving channel. When the Re is 3942, the flow is turbulent, and the PCA significantly decreases. The minimum value is 17.2% in the channel with 45° bends. The linear fitting model has a good correlation between PCA and Re, and R^2 values range from 0.96 to 0.98. These trends indicate that the smaller bending angle has more effect on particle deposition than the straight channel.

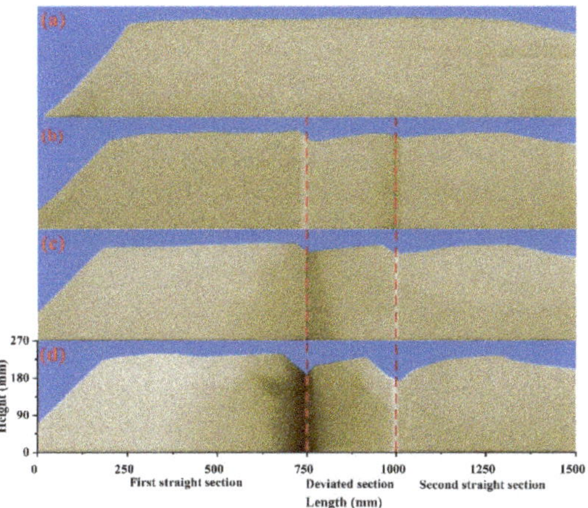

Figure 12. Snapshots of 40/70 mesh sand dunes in four channels at the stable status at Re = 946, S = 0.05, and $L_1/L = 0.5$, front view: (**a**) The straight channel, θ = 180°; (**b**) The curving channel, θ = 135°; (**c**) The curving channel, θ = 90°; (**d**) The curving channel, θ = 45°.

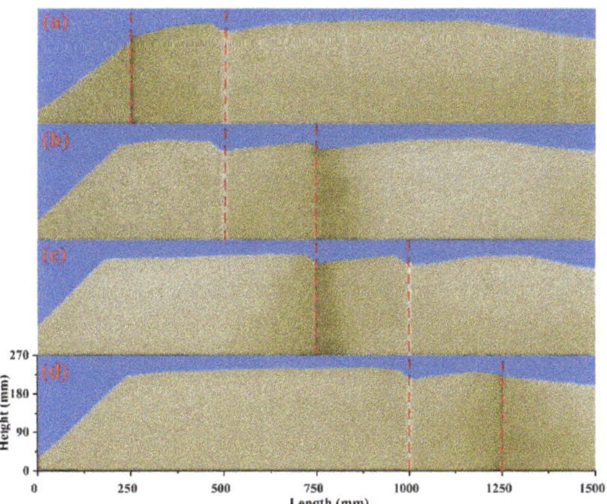

Figure 13. Snapshots of 40/70 mesh sand dunes in four channels at the stable status at Re = 946, S = 0.05 and θ = 90°, front view: (**a**) $L_1/L = 0.16$; (**b**) $L_1/L = 0.33$; (**c**) $L_1/L = 0.5$; (**d**) $L_1/L = 0.66$.

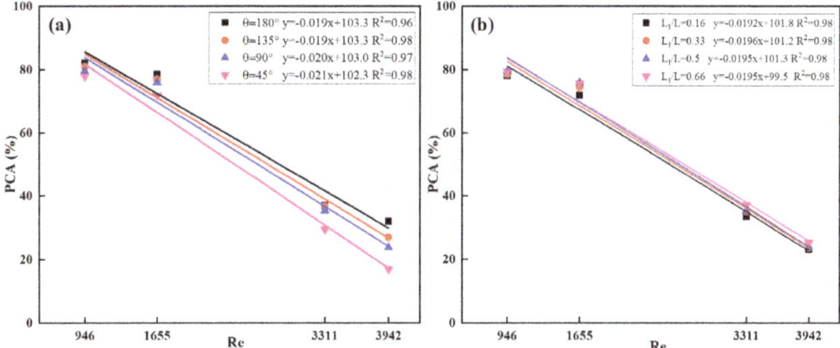

Figure 14. The correlations between Re and PCA: (**a**) The bending angle; (**b**) The distance ratio.

3.4. Effect of the Shields Number, S

The Shields number has a substantial impact on the particle placement in different channels. In Figure 15, two red dash lines in the figure mark the boundary of the curving section. In Figure 15a, a lot of sand is deposited in the straight channel to form a large dune. As the sand is transported in the curving channel with the angle of 135°, more sands flow out of the channel, and the dune area decreases, as shown in Figure 15b. In Figure 15d, the injected 70/140 mesh sands mainly deposit in the second straight section, and a few particles deposit in the curving section. The probable reason is that the curving channel with small bending angles induces turbulent flow, transporting more smaller particles out of the channel.

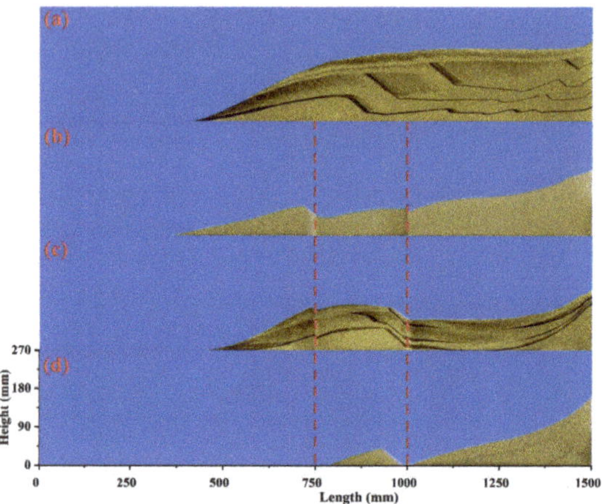

Figure 15. Snapshots of 70/140 mesh sand dunes in four channels at the stable status at Re = 3311 and $L_1/L = 0.5$, front view: (**a**) The straight channel, θ = 180°; (**b**) The curving channel, θ = 135°; (c) The curving channel, θ = 90°; (**d**) The curving channel, θ = 45°.

Figure 16 shows the front view of 70/140 mesh sand distributions in four channels for Case 10 (Table 2). When the ratio is 0.16, sands build up a dune in the second straight section, far away from the second bends. As the ratio increases to 0.66, particles build up a dune closer to the entrance due to the hindering effect, as shown in Figure 16d.

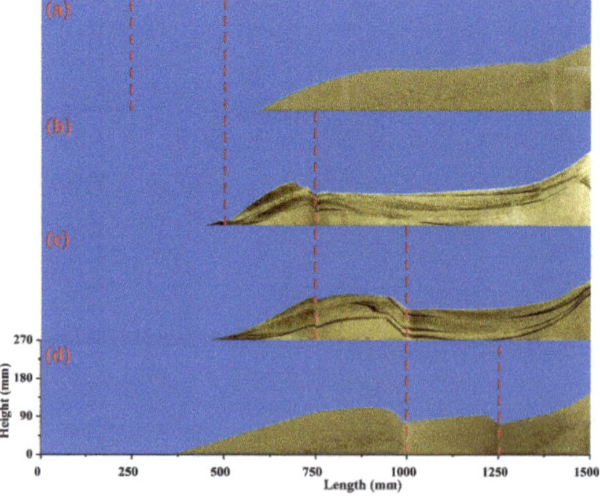

Figure 16. Snapshots of 70/140 mesh sand dunes in four channels at the stable status at *Re* = 3311 and θ = 90°, front view: (**a**) $L_1/L = 0.16$; (**b**) $L_1/L = 0.33$; (c) $L_1/L = 0.5$; (**d**) $L_1/L = 0.66$.

Figure 17 shows that the S has a negative linear correlation with PCA. The experiments are Cases 1 to 2 and Cases 9 to 14. The larger the Shields number, the lower the area of the dune. The minimum value is 8.5% in the channel with 45° bends. The linear fitting model has a good correlation between PCA and S, and R^2 values range from 0.87 to 0.98. Thus, it can conclude that the smaller particles smoothly flow through the curving section deeper into the channel. For the complex fracture system, the small particle is recommended for propping the remote fracture region.

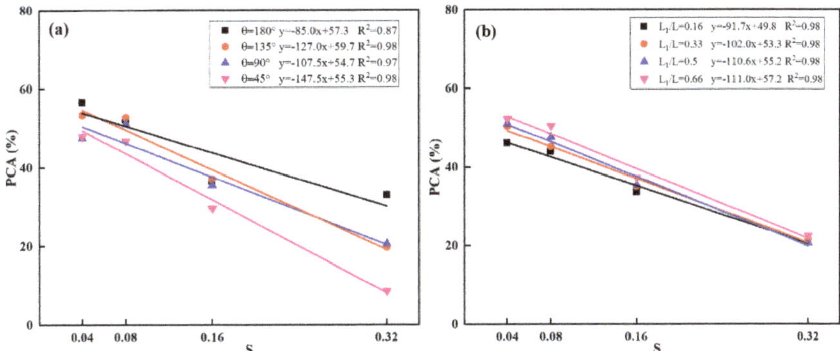

Figure 17. The correlations between S and *PCA*: (**a**) The bending angle; (**b**) The distance ratio.

4. Discussion

It is found that the *PCA* has relationships with four factors. According to the 56 experimental data, the correlation between the *PCA* and dimensionless parameters is modeled by the multivariable linear regression method, as shown in Equation (6).

$$PCA = 78.6 - 109.2S - 0.014Re + \ln\left(\theta^{5.5}\left(\frac{L_1}{L}\right)^{2.3}\right) \ R^2 = 0.98 \tag{6}$$

To evaluate the accuracy and reliability of developed models, the absolute percentage error (*APE*), mean absolute percentage error (*MAPE*), relative percentage error (*RPE*), and mean relative percentage error (*MRPE*) are used as follows:

$$APE = \left|\frac{E_i - P_i}{E_i}\right| \times 100\% \tag{7}$$

$$MAPE = \frac{1}{n}\sum_{i=1}^{n}\left|\frac{E_i - P_i}{E_i}\right| \times 100\% \tag{8}$$

$$RPE = |E_i - P_i| \times 100\% \tag{9}$$

$$MRPE = \frac{1}{n}\sum_{i=1}^{n}|E_i - P_i| \times 100\% \tag{10}$$

where E_i is the experimental result; P_i is the prediction value; n is the sample number.

Figure 18 shows the errors. In Figure 18a, *MRPE* is 2.77%, and the maximum *RPE* is 5.56%. In Figure 18b, *MAPE* is 6.34% for all experimental results, and the *APE* is 12.9%. The small errors mean the regression model can reliably predict the covered area of the particle dune in curved fractures.

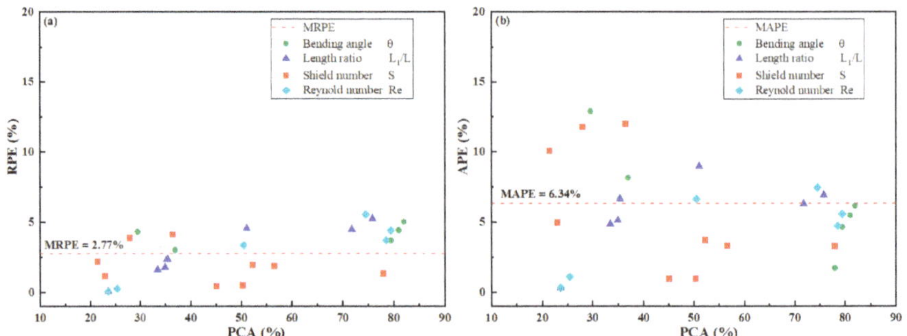

Figure 18. Error analysis. (**a**) Relative error; (**b**) Absolute error.

5. Conclusions

This work conducted a proppant transport experiment to reveal the particle placement characteristics in complex curving fractures. The effects of the fracture angle, the distance between the curving section and inlet, fluid velocity, and particle size on proppant placement were studied. Furthermore, the particle size distribution and proppant dune area were analyzed. Then, the primary conclusions are obtained as follows.

1. Due to the influence of the curving section, the liquid flow direction changes twice, leading to a change in particle settling behavior. The uneven deposition along the whole channel causes an irregular proppant dune. The large sand is prone to deposit at the front side of the dune, and small particles can move deeper into the curving fracture.
2. The curving section has a hindering effect on sand placement. The dune in the curving fracture is closer to the inlet compared to the straight fracture. With the decrease in the bending angle, more large particles would be transported out of the channel, and a lower and smaller dune is built up in the fracture.
3. As the curving section is farther away from the inlet, particles deposit near the inlet and form a larger proppant dune. In contrast, more particles flow through the curving section and deeper into the fracture. A large depleted zone is formed in the first and curving sections, significantly decreasing the fracture conductivity.
4. The Reynolds number and Shields number can significantly affect particle–fluid flow in curving fractures. The increase in Reynolds number and Shields number would significantly improve particle transport capacity, leading to the reduction of the covered area of the particle dune.
5. The proposed model, including four factors, can quantitatively predict the covered area of the proppant dune in curving fractures. The minor errors mean the high reliability of the equation.

Author Contributions: Conceptualization, Z.W.; methodology and formal analysis C.W.; writing—original draft preparation, L.Z. All authors have read and agreed to the published version of the manuscript.

Funding: This work is supported by the National Natural Science Foundation of China (Grant No. 52274035).

Institutional Review Board Statement: Not applicable.

Informed Consent Statement: Not applicable.

Data Availability Statement: The data that support the findings of this study are available from the corresponding author upon reasonable request.

Conflicts of Interest: The authors declare no conflict of interest.

References

1. Qu, H.; Tang, S.; Sheng, M.; Liu, Z.; Wang, R.; Hu, Y. Experimental investigation of the damage characteristics and breaking process of shale by abrasive waterjet impact. *J. Pet. Sci. Eng.* **2022**, *211*, 110165. [CrossRef]
2. Qu, H.; Tang, S.; Liu, Z.; McLennan, J.; Wang, R. Experimental investigation of proppant particles transport in a tortuous fracture. *Powder Technol.* **2021**, *382*, 95–106. [CrossRef]
3. Li, J.; Liu, P.; Kuang, S.; Yu, A. Visual lab tests: Proppant transportation in a 3D printed vertical hydraulic fracture with two-sided rough surfaces. *J. Pet. Sci. Eng.* **2021**, *196*, 107738. [CrossRef]
4. Qu, H.; Wang, R.; Ao, X.; Xue, L.; Liu, Z.; Lin, H. The Investigation of Proppant Particle-Fluid Flow in the Vertical Fracture with a Contracted Aperture. *SPE J.* **2022**, *27*, 274–291. [CrossRef]
5. Qu, H.; Liu, Y.; Zeng, S.; Xiao, H.; Lu, Y.; Liu, Z. Effect of Natural Fracture Characteristics on Proppant Transport and Placement in an Irregular, Nonplanar Fracture. *SPE J.* **2022**, *27*, 2208–2225. [CrossRef]
6. Warpinski, N.R.; Teufel, L.W. Influence of Geologic Discontinuities on Hydraulic Fracture Propagation (includes associated papers 17011 and 17074). *J. Pet. Technol.* **1987**, *39*, 209–220. [CrossRef]
7. Lee, H.; Olson, J.; Holder, J.; Gale, J.; Myers, R. The interaction of propagating opening mode fractures with preexisting discontinuities in shale: Shale vein-fracture interaction. *J. Geophys. Res. Solid Earth* **2014**, *120*, 169–181. [CrossRef]
8. Fu, W.; Savitski, A.A.; Damjanac, B.; Bunger, A.P. Three-dimensional lattice simulation of hydraulic fracture interaction with natural fractures. *Comput. Geotech.* **2019**, *107*, 214–234. [CrossRef]
9. Wu, X.; Huang, Z.; Cheng, Z.; Zhang, S.; Song, H.; Zhao, X. Effects of cyclic heating and LN2-cooling on the physical and mechanical properties of granite. *Appl. Therm. Eng.* **2019**, *156*, 99–110. [CrossRef]
10. Wu, X.; Huang, Z.; Zhang, S.; Cheng, Z.; Li, R.; Song, H.; Wen, H.; Huang, P. Damage Analysis of High-Temperature Rocks Subjected to LN2 Thermal Shock. *Rock Mech. Rock Eng.* **2019**, *52*, 2585–2603. [CrossRef]
11. Liou, T.-M.; Tzeng, Y.-Y.; Chen, C.-C. Fluid Flow in a 180 deg Sharp Turning Duct with Different Divider Thicknesses. *J. Turbomach.* **1999**, *121*, 569–576. [CrossRef]
12. Zhang, Y.; Chai, J. Effect of surface morphology on fluid flow in rough fractures: A review. *J. Nat. Gas Sci. Eng.* **2020**, *79*, 103343. [CrossRef]
13. Sahai, R.; Moghanloo, R.G. Proppant transport in complex fracture networks—A review. *J. Pet. Sci. Eng.* **2019**, *182*, 106199. [CrossRef]
14. Kern, L.; Perkins, T.; Wyant, R. The mechanics of sand movement in fracturing. *J. Pet. Technol.* **1959**, *11*, 55–57. [CrossRef]
15. Babcock, R.; Prokop, C.; Kehle, R. Distribution of propping agent in vertical fractures. In *Drilling and Production Practice*; American Petroleum Institute: New York, NY, USA, 1967.
16. Wang, J.; Joseph, D.D.; Patankar, N.A.; Conway, M.; Barree, R.D. Bi-power law correlations for sediment transport in pressure driven channel flows. *Int. J. Multiph. Flow* **2003**, *29*, 475–494. [CrossRef]
17. Woodworth, T.R.; Miskimins, J.L. Extrapolation of Laboratory Proppant Placement Behavior to the Field in Slickwater Fracturing Applications. In Proceedings of the SPE Hydraulic Fracturing Technology Conference, College Station, TX, USA, 29–31 January 2007; Society of Petroleum Engineers: College Station, TX, USA, 2007; p. 12.
18. Hu, X.; Wu, K.; Li, G.; Tang, J.; Shen, Z. Effect of proppant addition schedule on the proppant distribution in a straight fracture for slickwater treatment. *J. Pet. Sci. Eng.* **2018**, *167*, 110–119. [CrossRef]
19. Tong, S.; Mohanty, K.K. Proppant transport study in fractures with intersections. *Fuel* **2016**, *181*, 463–477. [CrossRef]
20. Fjaestad, D.; Tomac, I. Experimental investigation of sand proppant particles flow and transport regimes through narrow slots. *Powder Technol.* **2019**, *343*, 495–511. [CrossRef]
21. Zeng, H.; Jin, Y.; Qu, H.; Lu, Y.-H. Experimental investigation and correlations for proppant distribution in narrow fractures of deep shale gas reservoirs. *Pet. Sci.* **2021**, *19*, 619–628. [CrossRef]
22. Tong, S.; Singh, R.; Mohanty, K.K. A visualization study of proppant transport in foam fracturing fluids. *J. Nat. Gas Sci. Eng.* **2018**, *52*, 235–247. [CrossRef]
23. Hou, L.; Sun, B.; Wang, Z.; Li, Q. Experimental study of particle settling in supercritical carbon dioxide. *J. Supercrit. Fluids* **2015**, *100*, 121–128. [CrossRef]
24. Wen, Q.; Wang, S.; Duan, X.; Li, Y.; Wang, F.; Jin, X. Experimental investigation of proppant settling in complex hydraulic-natural fracture system in shale reservoirs. *J. Nat. Gas Sci. Eng.* **2016**, *33*, 70–80. [CrossRef]
25. Sahai, R.; Miskimins, J.L.; Olson, K.E. Laboratory Results of Proppant Transport in Complex Fracture Systems. In Proceedings of the SPE Hydraulic Fracturing Technology Conference, The Woodlands, TX, USA, 4–6 February 2014; Society of Petroleum Engineers: The Woodlands, TX, USA, 2014; p. 26.
26. Li, N.Y.; Li, J.; Zhao, L.Q.; Luo, Z.F.; Liu, P.L.; Guo, Y.J. Laboratory Testing on Proppant Transport in Complex-Fracture Systems. *SPE Prod. Oper.* **2017**, *32*, 382–391. [CrossRef]
27. Fan, L.F.; Wang, H.D.; Wu, Z.J.; Zhao, S.H. Effects of angle patterns at fracture intersections on fluid flow nonlinearity and outlet flow rate distribution at high Reynolds numbers. *Int. J. Rock Mech. Min. Sci.* **2019**, *124*, 104136. [CrossRef]
28. Wang, X.; Yao, J.; Gong, L.; Sun, H.; Liu, W. Numerical simulations of proppant deposition and transport characteristics in hydraulic fractures and fracture networks. *J. Pet. Sci. Eng.* **2019**, *183*, 106401. [CrossRef]
29. Yue, M.; Zhang, Q.; Zhu, W.; Zhang, L.; Song, H.; Li, J. Effects of proppant distribution in fracture networks on horizontal well performance. *J. Pet. Sci. Eng.* **2020**, *187*, 106816. [CrossRef]

30. Miskimins, J. *Hydraulic Fracturing: Fundamentals and Advancements*; Society of Petroleum Engineers: Danbury, CT, USA, 2020.
31. Fernández, M.E.; Sánchez, M.; Pugnaloni, L.A. Proppant transport in a scaled vertical planar fracture: Vorticity and dune placement. *J. Pet. Sci. Eng.* **2019**, *173*, 1382–1389. [CrossRef]
32. Chhabra, R.P. *Bubbles, Drops, and Particles in Non-Newtonian Fluids*, 2nd ed.; CRC Press: Boca Raton, FL, USA, 2007.
33. Zohuri, B. *Dimensional Analysis and Self-Similarity Methods for Engineers and Scientists*; Springer: Berlin/Heidelberg, Germany, 2015.
34. Mack, M.; Sun, J.; Khadilkar, C. Quantifying Proppant Transport in Thin Fluids: Theory and Experiments. In Proceedings of the SPE Hydraulic Fracturing Technology Conference, The Woodlands, TX, USA, 4–6 February 2014; Society of Petroleum Engineers: The Woodlands, TX, USA, 2014; p. 14.
35. Morsi, S.A.; Alexander, A.J. An investigation of particle trajectories in two-phase flow systems. *J. Fluid Mech.* **1972**, *55*, 193–208. [CrossRef]
36. Miller, M.C.; McCave, I.N.; Komar, P.D. Threshold of sediment motion under unidirectional currents. *Sedimentology* **1977**, *24*, 507–527. [CrossRef]
37. McClure, M. Bed load proppant transport during slickwater hydraulic fracturing: Insights from comparisons between published laboratory data and correlations for sediment and pipeline slurry transport. *J. Pet. Sci. Eng.* **2018**, *161*, 599–610. [CrossRef]
38. Liu, Y. Settling and Hydrodynamic Retardation of Proppants in Hydraulic Fractures. Ph.D. Thesis, University of Texas at Austin, Austin, TX, USA, 2006.
39. Joseph, H.; Spurk, N.A. *Fluid Mechanics*; Springer: Berlin/Heidelberg, Germany, 2020.
40. Anschutz, D.A.; Lowrey, T.A.; Stribling, M.; Wildt, P.J. An In-Depth Study of Proppant Transport and Placement with Various Fracturing Fluids. In Proceedings of the SPE Annual Technical Conference and Exhibition, Calgary, AB, Canada, 30 September–2 October 2019; Society of Petroleum Engineers: Calgary, AB, Canada, 2019; p. 27.

 energies

MDPI

Article

Investigation on the Propagation Mechanisms of a Hydraulic Fracture in Glutenite Reservoirs Using DEM

Jing Tang [1], Bingjie Liu [2] and Guodong Zhang [3,*]

1 CNOOC Research Institute Ltd., Beijing 100028, China
2 Department of Energy and Power Engineering, Tsinghua University, Beijing 100190, China
3 College of Electromechanical Engineering, Qingdao University, Qingdao 266061, China
* Correspondence: zhangguodong@qust.edu.cn

Abstract: The geometry heterogeneity induced by embedded gravel can cause severe stress heterogeneity and strength heterogeneity in glutenite reservoirs, and subsequently affect the initiation and propagation of hydraulic fractures. Since the discrete element method (DEM) can accurately describe the inter-particle interactions, the macromechanical behavior of glutenite specimen can be preciously represented by DEM. Therefore, the initiation and propagation mechanisms of hydraulic fractures were investigated using a coupling seepage-DEM approach, the terminal fracture morphologies of hydraulic fractures were determined, and the effects of stress differences, permeability, and gravel strength were studied. The results show that the initiation and propagation of hydraulic fractures are significantly influenced by embedded gravel. In addition, the stress heterogeneity and strength heterogeneity induced by the gravel embedded near the wellbore increase local initiation points, causing a complicated fracture network nearby. Moreover, due to the effect of local stress heterogeneity, gravel strength, and energy concentration near the fracture tip, four interactions of attraction, deflection, penetration, and termination between propagating fractures and encountering gravel were observed.

Keywords: hydraulic fracturing; glutenite; DEM; stress heterogeneity; fracture propagation

Citation: Tang, J.; Liu, B.; Zhang, G. Investigation on the Propagation Mechanisms of a Hydraulic Fracture in Glutenite Reservoirs Using DEM. *Energies* **2022**, *15*, 7709. https://doi.org/10.3390/en15207709

Academic Editor: Daoyi Zhu

Received: 3 August 2022
Accepted: 10 October 2022
Published: 19 October 2022

1. Introduction

In recent years, the utilization of oil and gas in the unconventional reservoirs [1,2], including gas shale [3,4], oil shale, and glutenite has attracted increased attention. Hydraulic fracturing is essential to increase production, which can form fractures with high conductivity, thus enhancing the ability of oil or gas to flow into the wells and increasing their productivity for cost-effective mining. Therefore, understanding the initiation mechanism and final patterns of fractures is of great significance for hydraulic fracturing design and treatment. However, due to severe geometric, stress, and strength heterogeneities, the initiation mechanism and propagation morphologies of hydraulic fractures are complicated [5].

Hydraulic fracturing involves complex seepage-stress-damage coupling processes. The geometric, stress, and strength heterogeneities inside the heterogeneous reservoirs seriously affect the fracture initiation and propagation of hydraulic fractures, and ultimately lead to complex fracture morphologies. Microseism, acoustic emission (AE), CT scanning, and nuclear magnetic resonance [6–8] have been widely used in the in situ monitoring and laboratory experiments to explore initiation and propagation laws of fractures. However, those techniques cannot explain the nucleation and evolution mechanism of fractures from the micro scale. Alternatively, numerical simulation technology can accurately reveal the micro-initiation and macro-propagation behaviors of hydraulic fractures [9]. In addition, the finite element method (FEM), boundary element method (BEM), and extended finite element method (XFEM) are broadly used. As the finite element cannot be re-meshed during simulation, only planar fractures are predicted [10]. Although BEM and XFEM can simulate the complex fracture morphologies, DEM cannot capture multiple simultaneous initiation points and the interaction between fractures in the propagation process [11].

In recent years, DEM has been broadly applied in the numerical simulation of hydraulic fracturing due to its unique advantages in describing the micromechanical response. Compared with continuous methods, DEM can accurately reveal the mechanics of microscopic particle by tracking and calculating particle interactions, in order that the nucleation, initiation, and propagation of hydraulic fractures can be truly reappeared. Meanwhile, DEM can also describe the complex fracture morphologies and accurately capture the multi-point initiation and propagation [12–14].

Glutenites are generally distributed in tight oil and gas reservoirs, and since glutenite reservoirs are usually deeply buried, their permeability and porosity are extremely low. For glutenite, the embedded gravel increases the complexity of initiation and propagation of hydraulic fractures in glutenite. True tri-axial experiments show that although the principal stress state dominates the propagation of hydraulic fractures in glutenite, gravel plays a key role in this process. When fractures encounter gravel, there are four interaction patterns: Attraction, deflection, penetration, and termination [15–17]. The geometric heterogeneity induced by gravel greatly adds the difficulty of numerical simulation of hydraulic fracturing in glutenite. The FSD model [5,18,19] is mainly used to characterize the influence of gravel on the initiation and propagation of hydraulic fractures. Although the true tri-axial experiment and numerical simulation explore the laws of fracture propagation in glutenite from different angles, it is still difficult to accurately characterize the rock mechanics behavior of glutenite and the seepage-stress-damage coupling problem involved in hydraulic fracturing.

Since DEM can directly represent the microstructure and mechanical characteristics of rock particles, the application of DEM to simulate the fracture initiation and propagation mechanism in glutenite is very adaptive and robust [20,21]. Nevertheless, few studies on DEM have been reported. Based on DEM, this paper explored the rock mechanics response from the micro scale, simulated the seepage process with the network flow model, and coupled the seepage with the discrete element process to investigate the microscopic mechanism of fracture initiation and propagation in glutenite. In addition, the effects of stress difference, permeability, gravel strength, injection rate, and well diameter were studied in this work.

2. Computational Models

2.1. Parallel Bond Model

The DEM simulates rock by bonding microparticles and characterizes the macromechanical behavior of the rock based on the interaction of microparticles. In this paper, a parallel bond model [22,23] is used to generate a numerical specimen of glutenite. After bonding, the particles can withstand normal force, shear force, and bending moment, which are calculated according to the following formula:

$$\overline{F}_n = \left(\overline{F}_n\right)_0 + \overline{k}_n \overline{A} \Delta\delta_n \quad (1)$$

$$\overline{F}_s = \left(\overline{F}_s\right)_0 + \overline{k}_s \overline{A} \Delta\delta_s \quad (2)$$

$$\overline{M}_t = \left(\overline{M}_t\right)_0 - \overline{k}_s \overline{J} \Delta\theta_t \quad (3)$$

where $\overline{F}_n$ is the normal force between particles (in N); $\left(\overline{F}_n\right)_0$ is the normal force at last time step (in N); $\overline{k}_n$ is the normal stiffness (in MPa/m); $\overline{A}$ is the bonding area between adjacent particles (in m^2); $\Delta\delta_n$ is the relative normal-displacement increment at the contact of adjacent particles (in m); $\overline{F}_s$ is the shear force between the particles (in N); $\left(\overline{F}_s\right)_0$ is the shear force at last time step (in N); $\overline{k}_s$ is the shear stiffness (in MPa/m); $\Delta\delta_s$ is the relative shear-displacement increment at the contact of adjacent particles (in m); $\overline{M}_t$ is the inter-particle bending moment (in N · m); $\left(\overline{M}_t\right)_0$ is the bending moment at last time step (in N · m); $\bar{J}$ is the polar moment of inertia at the contact of adjacent particles (in m^4); and $\Delta\theta_t$ is the relative twist-rotation increment (°).

2.2. Failure Criterions

DEM can accurately represent the mechanical interaction between microparticles. According to the bonding relationship between the particles, it can be judged whether cracks are generated. In other words, when the tensile stress between the particles exceeds the tensile strength limit, the normal bonding between the two particles is broken, thereby forming tensile micro-crack; when the shear stress between the particles exceeds the shear strength limit, the two particles relatively slip, and shear micro-crack is formed. Therefore, the initiation criterion can be expressed by the following formula [21,24], and the micro-crack whose direction is perpendicular to the particle contact direction and length is equal to the average of the adjacent particle sizes, is located at the failure of the bonding:

$$\sigma_t > pb\_ten \tag{4}$$

$$\tau > \sigma_n \tan\varphi + pb\_coh \tag{5}$$

where σ_t is the tensile stress (in MPa); *pb_ten* is the tensile strength limit (in MPa); τ is the shear stress (in MPa); σ_n is the normal compressive stress (in MPa); φ is the internal friction angle (°); and *pb_coh* is the cohesive (in MPa).

2.3. Implementation of Flow-DEM Coupling

Seepage involves a complex flow-stress coupling process. This paper simulates the seepage of fracturing fluid in porous media based on the modified network flow model [24] and couples it with the stress module characterized by discrete elements. Adjacent contact particles enclose a domain (Figure 1, green lines), and neighboring domains are connected by channels whose positions are at the contact of adjacent particles and directions are perpendicular to the contact direction between the particles, as shown in Figure 1. Due to the pressure difference between adjacent domains, the fluid flows from one domain to another, causing its pressure rises. Accordingly, the particles start to move under the influence of the increased force, leading to the changes in local stress, which in turn changes the volume of the domain and its internal fluid pressure, and further affects the fluid flow. The fluid flow through the flow tube can be calculated according to the Hagen-Poiseuille formula [25]:

$$Q = \frac{w^3}{12\mu}\frac{\Delta p}{L} \tag{6}$$

where Q is the volume flow through the channel (in m^3/s); w is the channel aperture (in m); μ is the fluid viscosity (in Pa · s); Δp is the pressure difference between two adjacent domains (in Pa); and L is the channel length (in m).

The amount of fluid pressure change in the domain can be calculated according to the following formula [26]:

$$\Delta p = \frac{K}{V_d}\left(\sum Q\Delta t - \Delta V_d\right) \tag{7}$$

where Δp is the pressure change caused by fluid exchange between the domains (in Pa); K is the fluid bulk modulus (in Pa); V_d is the domain volume (in m^3); Δt is the time step (in s); and ΔV_d is the volume change of the domain (in m^3).

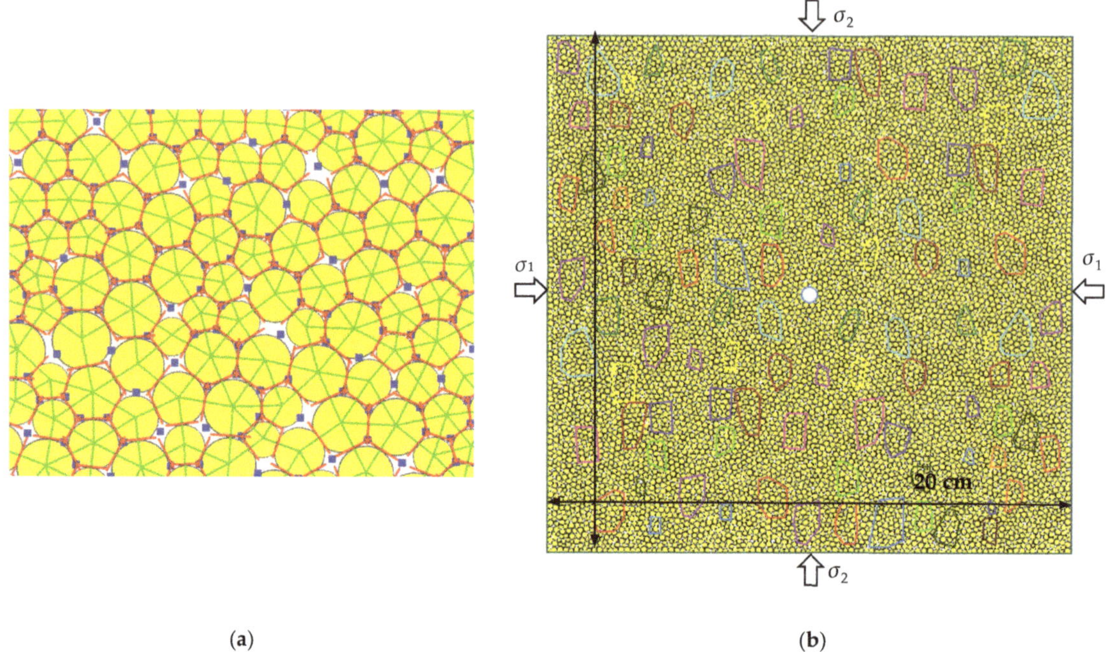

Figure 1. Schematic diagram of fluid and glutenite numerical models. (**a**) Network flow model. (**b**) Boundary conditions.

The aperture of the channel is closely related to the bond between particles. In essence, the seepage-stress coupling causes the particles to move, which in turn changes the aperture of the channel. When the bond is broken, the channel aperture at the contact is equal to the actual distance between the particles. Meanwhile, when the particles are compressed, the channel aperture can be calculated according to the residual aperture (the width at which the particles only contact), which can be expressed as follows [24]:

$$w = \frac{w_0 F_0}{F + F_0} \tag{8}$$

where w is the aperture of the channel under the compression of the particles (in m); w_0 is the residual aperture of the channel (in m); F_0 is the contact force of the particle when the aperture of the channel is equal to half of the residual aperture (in N); and F is the normal contact force of the particle (in N).

2.4. Simulation Scheme

Figure 1b shows the numerical model and boundary conditions of the glutenite. Considering the calculation cost, a laboratory-scale numerical model (20 × 20 cm) was built, which was filled with 10,394 particles, and adjacent particles were bonded to form rock through the parallel bond model. The particle radii were 0.6 mm $\leq R \leq$ 1.5 mm, according to the Rosin-Rammler distribution [27]. The gravel was created by the cutting method. In this method, polygons with different sizes, shapes, and positions were generated randomly and the particles surrounded by the polygons constitute gravel which was endowed with high mechanical parameters. A wellbore with a radius of 3 mm was established in the center of the model, and the boundary of constant injection rate was set to the wellbore. According to the servo principle, the external boundary motion of the glutenite was controlled to add confining pressure to the numerical model, and the horizontal principal stresses were set to σ_1 and σ_2, respectively.

In this work, the microscopic parameters used in the DEM simulation refer to those calibrated by Ju et al. [21], as shown in Table 1. The confining stresses are applied on the boundaries of the model under a servo-mechanism using four movable walls surrounding the rock specimen. The primary far-field horizontal geo-stress in x-direction is defined as σ_1, while σ_2 refers to the secondary far-field horizontal geo-stress in y-direction. In addition, the validation of DEM approach on the simulation of fracture propagation in glutenite reservoirs was carried out in our previous work [26], and a good agreement was observed. Moreover, as a typical DEM solver, particle flow code (PFC2D) was applied to implement the simulation, while the seepage module was self-defined in the code.

Table 1. Simulation parameters.

Model Properties	Value
Particle density (kg/m^3)	2600
Young's modulus of matrix (GPa)	22.29
Young's modulus of gravel (GPa)	43.8
Friction coefficient of particles in matrix	0.57
Friction coefficient of particles in gravel	0.57
Friction angle of particles in matrix (°)	26.13
Friction angle of particles in gravel (°)	50
Shear/normal spring stiffness ratio	2.5
Contact-bond tensile strength of matrix (MPa)	15
Contact-bond tensile strength of gravel (MPa)	25
Contact-bond cohesion of matrix (MPa)	52.44
Contact-bond cohesion of gravel (MPa)	100
Bulk modulus of injection fluid (GPa)	2.0
Fluid viscosity ($mPa \cdot s$)	1.0

The geo-stress state is generally considered to control the direction of hydraulic fracture propagation. In accordance with the linear elastic theory, the hydraulic fracture mainly extends along the direction of the maximum principal stress, thus forming a bi-wing fracture [5]. However, for glutenite reservoirs, the presence of gravel leads to severe geometric and stress heterogeneities, affecting the initiation and propagation of fractures. To clarify this effect, the hydraulic fracturing processes in glutenite under four kinds of geo-stress differences were simulated. The applied geo-stress states are shown in Table 2. In addition, four permeabilities of 0.0005×10^{-3} 0.005×10^{-3}, 0.05×10^{-3}, and 0.1×10^{-3} μm^2 were applied to evaluate its influence on the initiation and propagation of hydraulic fractures. The matrix-gravel strength difference would affect the interaction between hydraulic fractures and encountering gravel, and subsequently change the terminal fracture morphologies, thus two gravel strength ratios of 1.6 and 1.8 were used. Moreover, the influence of injection rate and well diameter was investigated, and two injection rates of 2×10^{-4} m^3/s and 4×10^{-4} m^3/s were used, while the initiation and propagation of hydraulic fractures under two well diameters (2 and 4 mm) were evaluated.

Table 2. Numerical model stress state setting table.

Principal Stress	Model 1	Model 2	Model 3	Model 4
σ_1/MPa	3	6	8	10
σ_2/MPa	5	5	5	5

3. Results and Discussion

3.1. Effect of Geo-Stress Difference

According to the circumferential stress theory [21], when the circumferential stress on the well wall is greater than the bond strength limit, the bond between the particles breaks, thereby forming micro-cracks, namely, crack initiation. However, due to the local stress heterogeneity induced by the geometrical heterogeneity near the wellbore, it is easy to form multiple initiation points at the well wall, and the number of initiation points increases as the principal stress difference decreases. Figure 2 shows the fracture propagation patterns under different principal stress differences. The results show that crack initiation near the wellbore wall is controlled by the stress heterogeneity and the principal stress state. The gravel around the wellbore leads to local stress concentration, which enhances the stress heterogeneity, thus affecting the initiation and propagation of fractures, and it is dominant in influencing the crack initiation. It is clear from Figure 2 that near the wellbore wall, the fracture mainly extends to the concentrated area of gravel, but its macro propagation direction is still controlled by the principal stress state. Near the wellbore wall, the injected fluid pressure and fracture spread synchronously. The energy of fracture initiation and propagation mainly comes from the fluid pressure [20], which is sufficient to form the multi-branch cracks and even the broken zones, as shown in Figure 2c,d. As the injection pressure increases, the crack gradually extends forward. At a distance from the wellbore region, the fluid pressure spread lags far behind the fracture propagation, thus the local stress concentration controls the initiation and propagation of fractures, allowing for the fractures to mainly extend along the direction of the maximum principal stress.

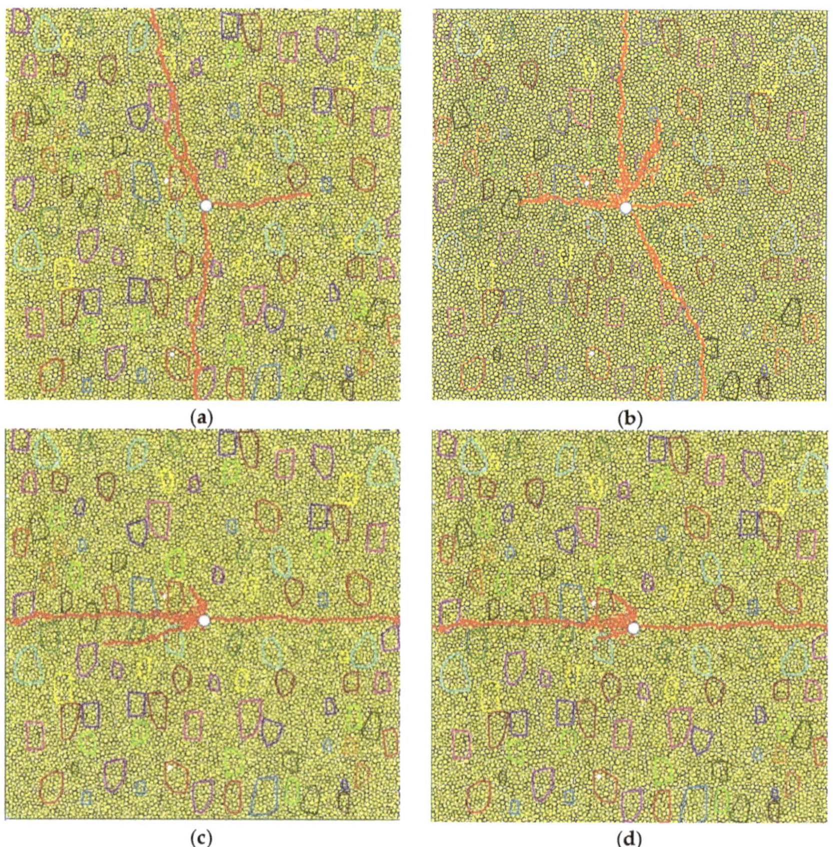

Figure 2. The fracture morphologies under different stress differences. (**a**) σ_1 = 3 MPa, σ_2 = 5 MPa. (**b**) σ_1 = 6 MPa, σ_2 = 5 MPa. (**c**) σ_1 = 8 MPa, σ_2 = 5 MPa. (**d**) σ_1 = 10 MPa, σ_2 = 5 MPa.

When the propagating fracture encounters gravel, a complex interaction between the two occurs, thus affecting the propagation direction and final morphology of the fracture. Initially, there is a serious stress concentration at the gravel boundary. When the fracture approaches the gravel, micro-cracks appear at the gravel boundary, thereby attracting the main fracture to the gravel boundary, namely, the attraction. Subsequently, the propagating fracture interacts with the gravel. If the gravel strength is low, the fracture can continue to penetrate through the gravel; if the gravel strength is high, the fracture will extend along the gravel boundary, resulting in deflection; if the energy at the tip of the fracture is insufficient, the fracture is trapped by the gravel and the termination appears, as shown in Figure 2b. In addition, the interaction between the fracture and the gravel is affected by the approximate relation of the fracture. When the fracture deviates from the gravel, it tends to be attracted by the gravel. Nevertheless, when the fracture hits the gravel on-head during propagation, the gravel can prevent the fracture from propagating further. Moreover, when the energy of the fracture tip is high, the fracture can deflect from or penetrate the gravel to continue extending.

Figure 3 shows the initiation and penetration times of fractures under different geo-stress differences. It is clear from the figure that as the stress difference increases, the initiation and penetration times are greatly reduced. When σ_1 = 3 MPa and σ_2 = 5 MPa, the gravel-induced stress concentration is mainly distributed in the direction of minimum principal stress, while the stress concentration caused by the principal stress is mainly along the direction of the maximum principal stress. Therefore, the injected fluid simultaneously percolates in multiple directions, which causes the injection pressure in the wellbore to increase at a slower speed, as shown in Figure 4, thus the fracture takes a long time to accumulate energy to initiate and propagate. As the principal stress difference increases, the range of the fluid percolation in the wellbore decreases, thereby accelerating the rate of pressure increase. It is clear from Figure 4 that as the stress difference increases, the wellbore injection pressure reaches the peak earlier, and the peak pressure is higher. During the fluid injection pressurization process, due to the frequent interaction of fractures and gravel, the injection pressure fluctuates and decreases rapidly after reaching the peak pressure. However, since the fracture propagates along the path of least resistance, even if the high-strength gravel blocks the fracture propagation, the hydraulic fracture can always find the minimum resistance path and continue propagating in the matrix, thus the initiation pressure is less affected by the stress difference.

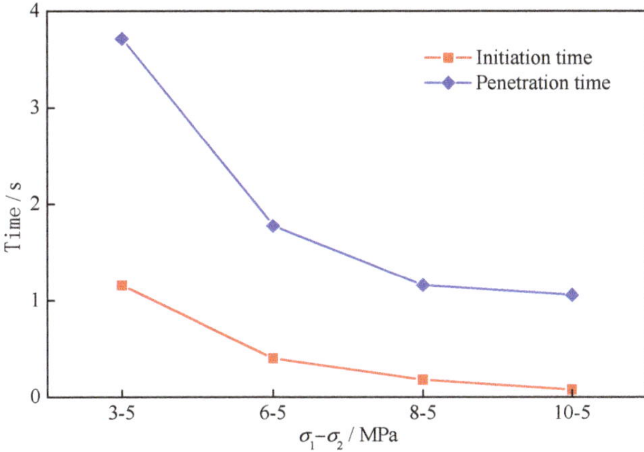

Figure 3. Initiation and penetration times of hydraulic fractures under different stress differences.

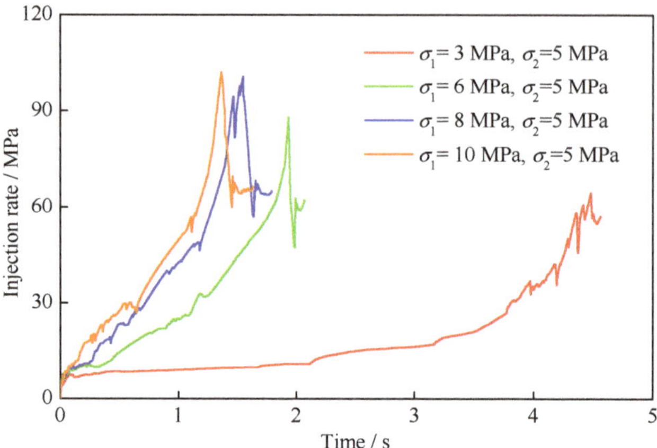

Figure 4. Injection pressure curves under different stress differences.

3.2. Effect of Permeability

The permeability can directly reflect the fluid percolation and pressure increase rates. Figure 5 shows the fracture propagation patterns under different permeabilities. When the permeability is small, it is difficult for the injection fluid to seep into the reservoir. Therefore, the fluid pressure in the wellbore rises rapidly, and the spread of the fluid pressure lags far behind, causing the surrounding stress to rapidly increase to the rock strength limit. Accordingly, the micro-cracks appear in a short time, as shown in Figure 6. Meanwhile, since the high energy accumulated near the wellbore is hard to release in a short time, the stresses of multiple points near the wellbore reach the strength limit synchronously, resulting in multiple initiation points and even broken zones, as shown in Figure 5a. Then, when the fracture encounters the gravel, it is more likely to extend inside the matrix since the energy spread is blocked, thus the deflection mainly occurs. As the permeability increases, more fluid can percolate into the reservoir, the pressure propagation rate of the fluid increases, and its effect for crack initiation starts to be dominant, resulting in a single initiation point near the wellbore. The greater the permeability, the more the fracture is inclined to extend in the direction of the maximum principal stress, forming a more regular fracture, namely, straighter bi-wing fracture, as shown in Figure 5d. The increase in permeability enhances the fluid percolation energy and reduces the rate of stress spread, thus the energy can be transmitted forward evenly. As a result, the hydraulic fracture has sufficient energy to penetrate the gravel to continue propagating, thus the penetration is increased. Additionally, high permeability reduces the energy concentration near the wellbore, thereby increasing the initiation and penetration times, as shown in Figure 6.

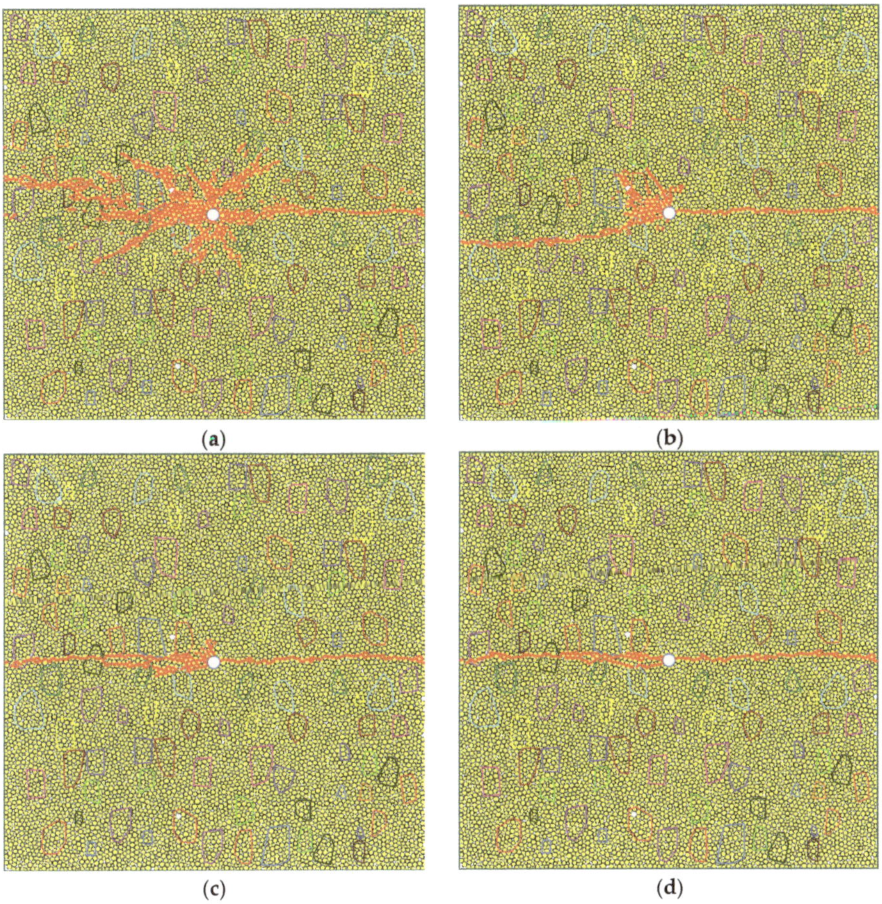

Figure 5. The fracture morphologies under different permeabilities. (**a**) $k = 0.0005 \times 10^{-3}$ μm^2. (**b**) $k = 0.005 \times 10^{-3}$ μm^2. (**c**) $k = 0.05 \times 10^{-3}$ μm^2. (**d**) $k = 0.1 \times 10^{-3}$ μm^2.

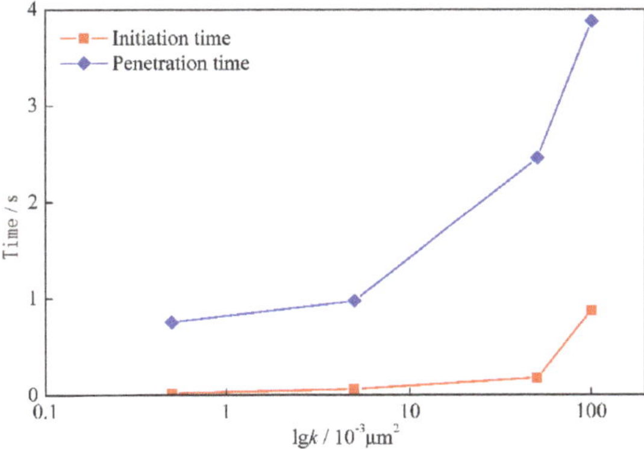

Figure 6. Initiation and penetration times of hydraulic fractures under different permeabilities.

3.3. Effect of Gravel Strength

Gravel strength mainly affects the interaction between the fractures and the gravel, which affects the initiation and propagation of fractures. Figure 7a,b shows the propagation patterns when the gravel strength is 1.6 and 1.8 times the initial strength, respectively. The results show that as the gravel strength increases, the penetration decreases. When the fracture encounters the gravel, although the gravel strength is significantly larger than the matrix, the hydraulic fracture can always find the least resistance channel in the matrix to extend, thus it mainly appears in the form of deflection. When the ratio of gravel strength to initial gravel strength is less than 1.6, the hydraulic fracture can easily penetrate the gravel, and the influence of gravel on the fracture propagation direction is not clear; when the strength ratio is higher ($\geq$1.8), the deflection occurs distinctly. After encountering the gravel, the fracture deflects multiple times along the boundary of the gravel, and at the same time forms a series of branch cracks, thus forming a complex fracture network. The greater the gravel strength, the more clear the fracture deflects, and the more complicated the final distribution morphology.

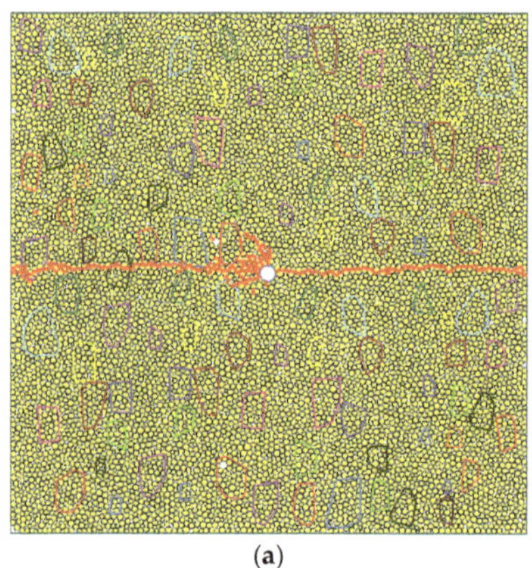
(a)

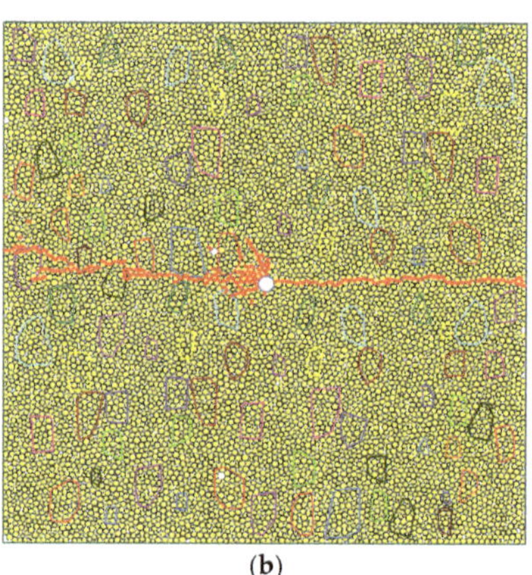
(b)

Figure 7. The fracture morphologies under different gravel strengths. (**a**) Gravel strength ratio is 1.6. (**b**) Gravel strength ratio is 1.8.

3.4. Effect of Injection Rate

The effect of injection rate on the initiation and propagation is similar to the effect of permeability. When the injection rate is high, the pressure in the wellbore increases rapidly due to the fact that the pressure spread lags behind, which causes energy concentration around the wellbore wall and multiple simultaneous initiation points, as shown in Figure 8; when injected at ultra-high speed, the wellbore wall even collapses and a broken zone is formed. At this time, the initiation and propagation are mainly affected by three factors: (1) Principal stress state; (2) random distribution of gravel and stress concentration caused by gravel irregularity; (3) energy concentration near the wellbore caused by high-speed injection, whose effect is mainly in the vicinity of the wellbore. In the early stage of hydraulic fracturing, the initiation and propagation of fractures are mainly affected by the latter two factors. As the fractures extend forward, the fluid pressure is difficult to spread to the far wellbore area in time, thus the initiation and propagation are mainly influenced by the principal stress. The fracture mainly extends along the direction of the maximum principal stress, thereby forming a regular bi-wing fracture. It is clear from Figure 8 that the local energy is difficult to release in time due to the high injection speed, thus the hydraulic fracture rapidly extends along the minimum resistance channel in the matrix. As a result, with the increase in the injection speed, the relationship between the fracture and gravel is traversed from penetration to deflection.

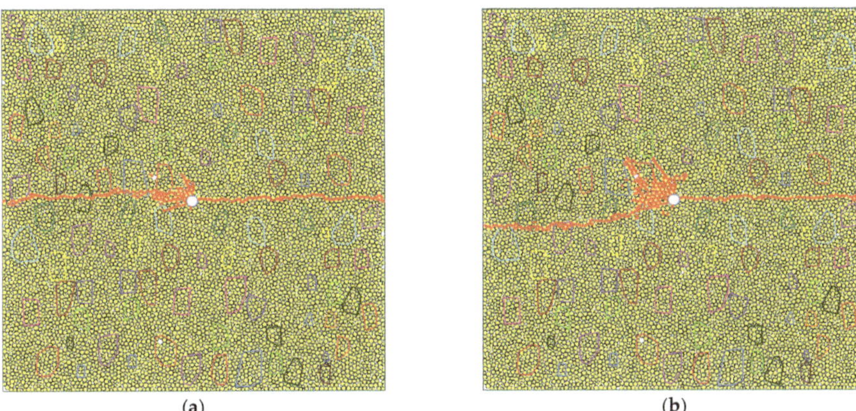

Figure 8. The fracture morphologies under different injection rates. (**a**) Injection rate is 2×10^{-4} m^3/s. (**b**) Injection rate is 4×10^{-4} m^3/s.

3.5. Effect of Well Diameter

The size of the wellbore seriously affects the surrounding geometric heterogeneity, causing stress heterogeneity, which in turn affects the initiation and propagation of fractures. Figure 9 shows the fracture propagation morphologies at different wellbore radii. The results show that when the diameter of the wellbore is small, the internal energy of the wellbore is sufficient, and it is mainly concentrated in the direction of the maximum principal stress. As a result, the fracture starts to initiate on the left and right sides of the wellbore and gradually extends outward, eventually forming a straight bi-wing shape. When the diameter of the wellbore is large, the heterogeneity around the wellbore is enhanced, and the energy distribution around the wellbore is dispersed, in order that multiple points reach the strength limit synchronously, leading to multi-point initiation. Moreover, the fracture slightly deflects from the horizontal direction when extending forward, which finally turns into a bi-wing shape with branched cracks. Compared with the large-size wellbore, the energy concentration is increased when the wellbore is small. At a distance from the wellbore there is still sufficient energy at the fracture tip to penetrate the gravel, while for the larger wellbore, the energy of the fracture tip is insufficient, it mainly extends forward in the form of deflection.

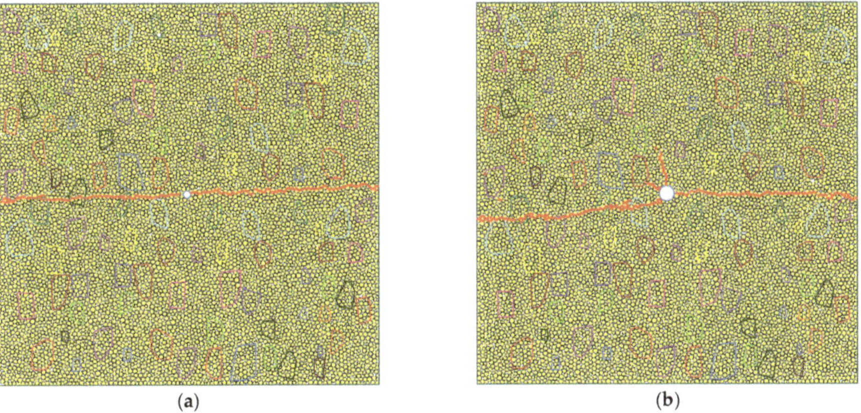

Figure 9. The fracture morphologies under different radii of the wellbore. (**a**) r_w = 2 mm. (**b**) r_w = 4 mm.

4. Discussion

For glutenite reservoirs, strong heterogeneity is induced by embedded gravel, and this causes the initiation and propagation of hydraulic fractures to be more complicated. It was reported that a substantial "shielding" effect on near fracture can be produced by embedded gravel, and this would result in a reduction in the energy release rate [28,29]. Therefore, the stress near the interfaces between gravel and matrix quickly increases as hydraulic fractures approach the gravel, and this change subsequently causes local failures, which gradually attract the propagating hydraulic fractures. It was found that a hydraulic fracture is anticipated to be attracted by the encountering gravel due to interface failures as it deviates from the centerline and propagates the gravel. However, when a fracture propagates along the centerline of the gravel, it may deflect or the gravel is penetrated depending on local stress distribution. Due to the limitation of true tri-axial experiments, a physical experiment on hydraulic fracturing in a glutenite specimen is rare, and the only one was presented by Ma et al. [15] to the best of our knowledge. Here, the authors found three interaction patterns between propagating fractures and encountering gravel, which were defined as attraction, deflection, and penetration [15]. Moreover, we observed these interaction patterns in this work, suggesting excellent robustness of the proposed approach on simulating this complicated initiation and propagation of hydraulic fractures. Furthermore, the interaction patterns are significantly influenced by permeability, gravel strength, and injection rate, ect., thus complicated morphologies of hydraulic fractures are usually anticipated in glutenite reservoirs.

Finally, geological stress controls the global propagation of hydraulic fracture, while local stress state significantly influences its initiation and propagation, and their combination gives rise to complicated fracture morphologies. The pore pressure is affected by the permeability and injection rate, which changes the local stress state, and is significantly influenced by gravel strength and well diameter, thus different fracture morphologies were observed under different parameters. In addition, compared with geological stress which is almost constant, local stress distribution can be changed by regulating the working parameters during the hydraulic fracturing treatment, thus these findings are helpful in increasing the stimulated reservoir volume in fields. However, since this work is only a laboratory scale simulation, the upscale of the simulation is required for better understanding of the mechanisms behind the initiation and propagation of hydraulic fracture in glutenite reservoirs.

5. Conclusions

(1) Based on the coupled seepage-DEM method, the mechanism of hydraulic fracture initiation and propagation in glutenite was explored, the interaction between fractures and gravel was revealed, and the effects of geo-stress difference, permeability, gravel strength, injection rate, and well diameter were analyzed.

(2) When hydraulic fractures encounter gravel, attraction, deflection, penetration, and termination are generally observed. The interaction mechanism between hydraulic fractures and gravel is affected by local stress heterogeneity, fracture tip energy, and gravel strength.

(3) The initiation of hydraulic fracture near the wellbore wall is controlled by local stress heterogeneity, while the existence of gravel increases the stress heterogeneity, which leads to simultaneous multi-point initiation. Moreover, local stress concentration dominates the propagation of hydraulic fractures, and when they enter into the far wellbore area, the influence of the geological stress state is further highlighted.

Author Contributions: Conceptualization, G.Z. and B.L.; methodology, G.Z.; formal analysis, G.Z.; data curation, G.Z.; writing—original draft preparation, G.Z. and J.T.; writing—review and editing, G.Z. and J.T. All authors have read and agreed to the published version of the manuscript.

Funding: This study was supported by the National Natural Science Foundation of China (Grant No. 51804175), Natural Science Foundation of Shandong Province, China (Grant No. ZR2018BEE005), and Fundamental Research Funds for the Central Universities.

Data Availability Statement: The data that support the findings of this study are available from the corresponding author upon reasonable request.

Conflicts of Interest: The authors declare no conflict of interest.

Nomenclature

Symbols	
F_n	normal force (N)
$(F_n)_0$	normal force at last time step (N)
k_n	normal stiffness (N/m)
$\overline{A}$	bonding area between adjacent particles (m^2)
$\overline{F}_s$	shear force between the particles (N)
$(F_s)_0$	shear force at last time step (N)
k_s	shear stiffness (N/m)
$\overline{M}_t$	inter-particle bending moment (N · m)
$(\overline{M}_t)_0$	bending moment at last time step (N · m)
$\overline{J}$	polar moment of inertia at the contact of adjacent particles (m^4)
pb_ten	tensile strength limit (MPa)
pb_coh	cohesion (MPa)
Q	volume flow through the channel (m^3/s)
dp	pressure difference between two adjacent domains (MPa)
L	channel length (m)
K	fluid bulk modulus (Pa)
V_d	domain volume (m^3)
w	channel aperture (m)
w_0	residual aperture (m)
F_0	normal force when the channel aperture reduces to half of its residual aperture (N)
F	normal contact force of the particle (N)
Greek Symbols	
$\Delta\delta_n$	relative normal-displacement increment (m)
$\Delta\delta_s$	relative shear-displacement increment (m)
$\Delta\theta_t$	relative twist-rotation increment (m)
σ_t	tensile stress (MPa)
τ	shear stress (MPa)
σ_n	normal compressive stress (MPa)
φ	internal friction angle (°)
μ	fluid viscosity (Pa · s)
Δp	pressure change caused by fluid exchange between the domains (Pa)
Δt	time step (s)
ΔV_d	volume change of the domain (m^3)

References

1. Liu, Z.; Bai, B.; Wang, Y.; Ding, Z.; Li, J.; Qu, H.; Jia, Z. Experimental study of friction reducer effect on dynamic and isotherm of methane desorption on Longmaxi shale. *Fuel* **2020**, *288*, 119733. [CrossRef]
2. Liu, Z.; Bai, B.; Tang, J.; Xiang, Z.; Zeng, S.; Qu, H. Investigation of Slickwater Effect on Permeability of Gas Shale from Longmaxi Formation. *Energy Fuels* **2021**, *35*, 3104–3111. [CrossRef]
3. Qu, H.; Tang, S.; Liu, Z.; Mclennan, J.; Wang, R. Experimental investigation of proppant particles transport in a tortuous fracture. *Powder Technol.* **2021**, *382*, 95–106. [CrossRef]
4. Qu, H.; Liu, Y.; Zeng, S.; Xiao, H.; Lu, Y.; Liu, Z. Effect of Natural Fracture Characteristics on Proppant Transport and Placement in an Irregular, Nonplanar Fracture. *SPE J.* **2022**, *27*, 2208–2225. [CrossRef]
5. Li, L.; Meng, Q.; Wang, S.; Li, G.; Tang, C. A numerical investigation of the hydraulic fracturing behaviour of conglomerate in Glutenite formation. *Acta Geotech.* **2013**, *8*, 597–618. [CrossRef]
6. Guo, C.; Xu, J.; Wei, M.; Jiang, R. Experimental study and numerical simulation of hydraulic fracturing tight sandstone reservoirs. *Fuel* **2015**, *159*, 334–344. [CrossRef]
7. Yushi, Z.; Shicheng, Z.; Tong, Z.; Xiang, Z.; Tiankui, G. Experimental Investigation into Hydraulic Fracture Network Propagation in Gas Shales Using CT Scanning Technology. *Rock Mech. Rock Eng.* **2016**, *49*, 33–45. [CrossRef]
8. Bunger, A.P.; Kear, J.; Dyskin, A.V.; Pasternak, E. Sustained acoustic emissions following tensile crack propagation in a crystalline rock. *Int. J. Fract.* **2015**, *193*, 87–98. [CrossRef]

9. Adachi, J.; Siebrits, E.; Peirce, A.; Desroches, J. Computer simulation of hydraulic fractures. *Int. J. Rock Mech. Min. Sci.* **2007**, *44*, 739–757. [CrossRef]
10. Chen, Z. Finite element modelling of viscosity-dominated hydraulic fractures. *J. Pet. Sci. Eng.* **2012**, *88*, 136–144. [CrossRef]
11. Eshiet, K.I.; Sheng, Y.; Ye, J. Microscopic modelling of the hydraulic fracturing process. *Environ. Earth Sci.* **2013**, *68*, 1169–1186. [CrossRef]
12. Al-Busaidi, A.; Hazzard, J.F.; Young, R.P. Distinct element modeling of hydraulically fractured Lac du Bonnet granite. *J. Geophys. Res. Earth Surf.* **2005**, *110*, 287–298. [CrossRef]
13. Zhou, J.; Zhang, L.; Pan, Z.; Han, Z. Numerical studies of interactions between hydraulic and natural fractures by Smooth Joint Model. *J. Nat. Gas Sci. Eng.* **2017**, *46*, 592–602. [CrossRef]
14. Fatahi, H.; Hossain, M.; Sarmadivaleh, M. Numerical and experimental investigation of the interaction of natural and propagated hydraulic fracture. *J. Nat. Gas Sci. Eng.* **2017**, *37*, 409–424. [CrossRef]
15. Ma, X.; Zou, Y.; Li, N.; Chen, M.; Zhang, Y.; Liu, Z. Experimental study on the mechanism of hydraulic fracture growth in a glutenite reservoir. *J. Struct. Geol.* **2017**, *97*, 37–47. [CrossRef]
16. Liu, P.; Ju, Y.; Ranjith, P.; Zheng, Z.; Chen, J. Experimental investigation of the effects of heterogeneity and geostress difference on the 3D growth and distribution of hydrofracturing cracks in unconventional reservoir rocks. *J. Nat. Gas Sci. Eng.* **2016**, *35*, 541–554. [CrossRef]
17. Li, N.; Zhang, S.; Ma, X.; Zou, Y.; Chen, M.; Li, S.; Zhang, Y. Experimental study on the propagation mechanism of hydraulic fracture in glutenite formations. *Chin. J. Rock Mech. Eng.* **2017**, *36*, 2383–2392. (In Chinese)
18. Rui, Z.; Guo, T.; Feng, Q.; Qu, Z.; Qi, N.; Gong, F. Influence of gravel on the propagation pattern of hydraulic fracture in the glutenite reservoir. *J. Pet. Sci. Eng.* **2018**, *165*, 627–639. [CrossRef]
19. Li, L.; Li, G.; Meng, Q.; Wang, H.; Wang, Z. Numerical simulation of propagation of hydraulic fractures in glutenite formation. *Rock Soil Mech.* **2013**, *34*, 1501–1507. (In Chinese)
20. Shimizu, H.; Murata, S.; Ishida, T. The distinct element analysis for hydraulic fracturing in hard rock considering fluid viscosity and particle size distribution. *Int. J. Rock Mech. Min. Sci.* **2011**, *48*, 712–727. [CrossRef]
21. Ju, Y.; Liu, P.; Chen, J.; Yang, Y.; Ranjith, P. CDEM-based analysis of the 3D initiation and propagation of hydrofracturing cracks in heterogeneous glutenites. *J. Nat. Gas Sci. Eng.* **2016**, *35*, 614–623. [CrossRef]
22. Zhang, X.P.; Wong, L.N.Y. Crack initiation, propagation and coalescence in rock-like material containing two flaws: A numerical study based on bonded-particle model approach. *Rock Mech. Rock Eng.* **2013**, *46*, 1001–1021. [CrossRef]
23. Potyondy, D.O.; Cundall, P. A bonded-particle model for rock. *Int. J. Rock Mech. Min. Sci.* **2004**, *41*, 1329–1364. [CrossRef]
24. Cundall, P. *Fluid Formulation for PFC2D*; Itasca Consulting Group: Minneapolis, Minnesota, 2000.
25. White, F. *Fluid Mechanics*, 7th ed.; McGraw-Hill: New York, NY, USA, 2009.
26. Zhang, G.; Sun, S.; Chao, K.; Niu, R.; Liu, B.; Li, Y.; Wang, F. Investigation of the nucleation, propagation and coalescence of hydraulic fractures in glutenite reservoirs using a coupled fluid flow-DEM approach. *Powder Technol.* **2019**, *354*, 301–313. [CrossRef]
27. RosIN, P. Laws governing the fineness of powdered coal. *J. Inst. Fuel* **1933**, *7*, 29–36.
28. Bush, M. The interaction between a crack and a particle cluster. *Int. J. Fract.* **1997**, *88*, 215–232. [CrossRef]
29. Li, R.; Chudnovsky, A. Variation of the energy release rate as a crack approaches and passes through an elastic inclusion. *Int. J. Fract.* **1993**, *59*, 69–74. [CrossRef]

Article

Abnormal Phenomena and Mathematical Model of Fluid Imbibition in a Capillary Tube

Wenquan Deng, Tianbo Liang *, Shuai Yuan, Fujian Zhou and Junjian Li

State Key Laboratory of Oil and Gas Resources and Prospecting, China University of Petroleum at Beijing, Beijing 102249, China; dwq10284413@163.com (W.D.); yuanshuai_2015@163.com (S.Y.); zhoufj@cup.edu.cn (F.Z.); junjian@cup.edu.cn (J.L.)

* Correspondence: liangtianboo@163.com

Abstract: At present, the imbibition behavior in tight rocks has been attracted increasing attention since spontaneous imbibition plays an important role in unconventional oil and gas development, such as increasing swept area and enhancing recovery rate. However, it is difficult to describe the imbibition behavior through imbibition experiment using tight rock core. To characterize the imbibition behavior, imbibition and drainage experiments were conducted among water, oil, and gas phases in a visible circular capillary tube. The whole imbibition process is monitored using a microfluidic platform equipped with a high frame rate camera. This study conducts two main imbibition experiments, namely liquid-displacing-air and water-displacing-oil experiments. The latter is a spontaneous imbibition that the lower-viscosity liquid displaces the higher-viscosity liquid. For the latter, the tendency of imbibition rate with time does not match with previous model. The experimental results indicate that it is unreasonable to take no account of the effect of accumulated liquid flowing out of the capillary tube on imbibition, especially in the imbibition experiments where the lower-viscosity liquid displaces the higher-viscosity liquid. A mathematical model is established by introducing an additional force to describe the imbibition behavior in capillary tube, and the model shows a good prediction effect on the tendency of imbibition rate with time. This study discovers and analyzes the effect of additional force on imbibition, and the analysis has significances to understand the imbibition behavior in tight rocks.

Keywords: spontaneous imbibition; hydraulic fracturing; microfluidics; tight rock

Citation: Deng, W.; Liang, T.; Yuan, S.; Zhou, F.; Li, J. Abnormal Phenomena and Mathematical Model of Fluid Imbibition in a Capillary Tube. *Energies* **2022**, *15*, 4994. https://doi.org/10.3390/en15144994

Academic Editor: Daoyi Zhu

Received: 22 May 2022
Accepted: 5 July 2022
Published: 8 July 2022

1. Introduction

At present, horizontal wells and hydraulic fracturing are two key technologies for developing tight oil and gas reservoirs [1,2]. However, due to both fast water breakthrough and production decline, the recovery rate of oil and gas is typically less than 10% [3,4]. In tight rocks, the imbibition behavior is strongly affected by the capillary force, where the pore throat size is less than 1 μm [5], and improving water imbibition in fracture and porous matrix can enhance the recovery rate after fracturing. In order to predict oil and water flow in small pores, it is necessary to establish a numerical solution of imbibition rate in the regular-shaped visual model, which considers the influence of viscosity, interfacial tension, and pore diameter on imbibition rate.

Imbibition behavior and its factors through laboratory experiments using single channels have been studied for a long time. Mathematical models have been proposed to describe the imbibition behavior, among which the most classical formulas are Washburn equation and Poiseuille equation [6]. As early as 1921, Lucas (1918) [7] and Washburn (1921) [8] mathematically described the capillary phenomenon in a vertically placed circular tube. In this study, the tube diameter was small enough to ignore the influence of gravity, and the Washburn equation was obtained on basis of Poiseuille equation under the condition that only viscous and capillary forces were considered. Similarly, the equation is consistent with that in a horizontal single channel. Washburn equation and Poiseuille equation are respectively shown as Equations (1) and (2).

$$L^2 = \frac{R\sigma \cdot cos\theta}{2\mu} \cdot t \quad (1)$$

$$q = \frac{\pi R^4 \Delta P}{8\mu \Delta L} \quad (2)$$

Therein, R represents radius of circular single channel; μ indicates the liquid viscosity; L represents the imbibition distance at the corresponding imbibition time t; θ is the contact angle; σ indicates surface tension; q denotes volume rate of flow; $\Delta P/\Delta L$ represents pressure gradient along the direction of the circular single channel.

Equation (1) not only shows a linear relationship between the imbibition distance and the square root of time, but also gives an idea to establish a mathematical model which describes the imbibition behavior that one liquid displaces another liquid. In the liquid-liquid imbibition, viscous force comes from the flow of displacing phase and displaced phase. Considering an extra applied pressure drop, Poiseuille mathematical model for liquid-liquid imbibition is shown as Equation (3) [6].

$$\frac{1}{8}\left(P_c + \Delta P_{applied}\right) \cdot R^2 t = \frac{1}{2}(\mu_w - \mu_{nw})L^2 + \mu_{nw} \cdot L_{channel} \cdot L \quad (3)$$

Therein, P_c represents capillary force; $\Delta P_{applied}$ indicates an extra applied pressure drop; μ_w and μ_{nw} represent the viscosity of wetting phase and non-wetting phase respectively; $L_{channel}$ is the length of circular single channel.

As shown in Equation (3), general studies of displacement in artificial core or micro model were conducted in the context that an extra pressure drop is applied [9,10]. However, spontaneous imbibition, which refers to the wetting phase flowing spontaneously into the porous media or single channel and displacing the non-wetting phase only by the capillary force [11], does not consider the extra pressure. Therefore, Equation (3) becomes Equation (4) after introducing the equation of capillary force.

$$\frac{\sigma R cos\theta}{4} t = \frac{1}{2}(\mu_w - \mu_{nw})L^2 + \mu_{nw} \cdot L_{channel} \cdot L \quad (4)$$

Equation (4) shows that the imbibition distance as a function of time is mainly related to the length and diameter of the single channel, contact angle, interfacial tension and wettability [12,13]. Wang et al. (2019) [12] proposed a semi-analytical mathematical model characterizing co-current spontaneous imbibition that water imbibes into oil-saturated tight sandstone, and found that the larger the maximum pore radius is, the higher the initial imbibition rate is.

Since this semi-analytical mathematical model is established on basis of a circular capillary tube model, the effect of radius on the imbibition rate is same to that in single-channel model. Hilpert (2009a; 2009b) [14,15] derived semi-analytical solutions to liquid imbibition into horizontal and inclined tubes which considers the dynamic contact angle based on the power law and series models, and these two semi-analytical solutions considered the gravity of liquid due to the large diameter of the circular tube.

Soares et al. (2005) [16] found that an uncertainty of contact angle in imbibition studies since the displaced liquid can stick to the capillary-tube wall and consequently influences contact angle during this imbibition process. The uncertainty of contact angle on process of imbibition is also related to interfacial tension [17,18]. Thus, the contact angle (which is influenced by the above factors) is likely to affect imbibition rate, but it does not affect the law of imbibition rate with time. In imbibition studies characterizing mathematically imbibition rate, experiments have generally been conducted in rectangular single channels due to the limit of etching techniques [19–21]. Therefore, in the imbibition experiment of rectangular single channels, the focus is how the length and width of the rectangular single channel influence the imbibition rate. The researchers developed different mathematical models based on their own experiments to characterize imbibition behavior in rectangular single channels with different sizes [21,22]. Yang et al. (2011, 2014) [23,24] studied the dynamics of capillary-driven flow in open hydrophilic microchannels, and found that the imbibition rate decreases with the increase of channel width. Chen (2014) [25] investigated the penetration of a wetting liquid into open metallic grooves and found that the imbibition rate is faster in deeper grooves. These basic studies provide a better understanding of different factors on imbibition behavior. However, an interesting and unconventional phenomenon was found in the study of Yang et al. (2014). When the lower-viscosity Deionized water (DI-water) displaces the higher-viscosity hexadecane, the imbibition rate shows a decreasing trend with time. This observation is contradictory to the trend proposed by Equation (4), and this is due to the effect of displaced phase on imbibition in single channel was not comprehensively considered.

In this study, imbibition and drainage experiments are conducted among water, oil and gas phases to understand the law of imbibition rate and evaluate the prediction effect of Washburn model and modified Poiseuille model. For imbibition experiment of water-displacing-oil, we modified the Poiseuille model and established this modified Poiseuille model on basis of Equation (4) by introducing a third force come from the reverse imbibition. This modified model explains the effect of displaced liquid that is displaced from single channel on the abnormal phenomenon in imbibition of liquid-displacing-liquid, and it also reveals a new concept that an additional force exists due to reverse imbibition outside the single channel when the lower-viscosity fluid displaces the higher-viscosity fluid. Furthermore, this study provides enlightenment for the analysis of the force during imbibition process in tight rocks.

2. Materials and Methods

2.1. Experimental Materials and Equipment

BZY-2 automatic interface tensioner is used for measuring the interfacial tension through the lifting-ring method. JY-PHB contact angle tester is used for the measurement of contact angle. The imbibition distance as a function of time were monitored using a microfluidic platform equipped with a fluorescent stereo microscope (Leica M165 FC), as shown in Figure 1.

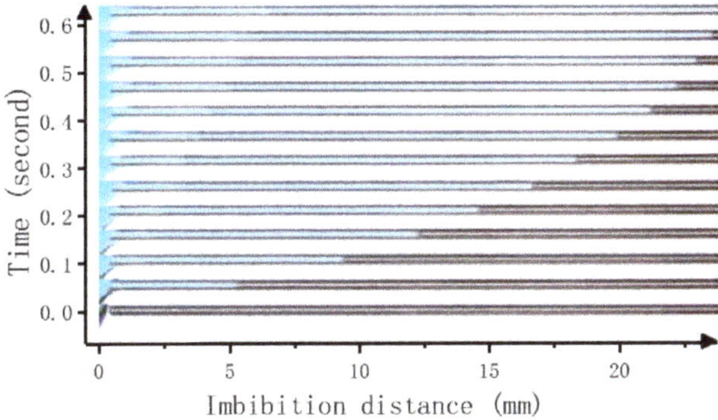

Figure 1. Imbibition that DI-water displaces air in one single channel.

The single-channel models of circular capillary tube used in this paper, which are made of fused silica, are 100 mm in length and 0.3, 0.4, 0.5, and 0.6 mm in the inner diameter. The single-channel models are hydrophilic, and different lengths can be designed for the single-channel model to explore the imbibition effect under different sizes.

The oil sample used is kerosene (from Chengdu Xinzhengtong Chemical Co., Ltd., Chengdu, China). Compared with other oil samples, kerosene does not affect the wettability of the channel wall during the spontaneous imbibition. The water phase is DI-water, which reduces the influence of impurities on spontaneous imbibition process.

2.2. Experimental Methods

Basic parameters are shown in Table 1. As shown in Table 1, the fluctuation of the contact angle between liquid (DI-water: 79.8° or kerosene: 67°) and air does not apparently influence its cosine. Besides, the contact angles in different tube diameters between liquid and air are almost the same with fluctuation less than 1.5°. However, in the experiment of DI-water-displacing-kerosene, with the range 82.3° to 87.9° of the contact angle between kerosene and DI-water in single channels with three different diameters. Although just ±1° fluctuation of contact angel occurs, greater changes are induced in the corresponding cosine, resulting in influence on the analysis of fitting. Accordingly, the four contact angles are not averaged.

Table 1. Property parameters.

Property	Value	
Viscosity of DI-water	1.0 mPa·s	
Viscosity of kerosene	2.2 mPa·s	
Surface tension between DI-water and air	66.5 mN/m	
Surface tension between kerosene and air	24.0 mN/m	
Interfacial tension between kerosene and DI-water	20.9 mN/m	
The contact Angle between DI-water and air in the single channel	79.8 ± 1.4°	
The contact Angle between kerosene and air in the single channel	67.0° ± 1.2	
The contact Angle between kerosene and DI-water in the single channel	0.3 × 10.60 mm	87.9°
	0.5 × 15.00 mm	86.2°
	0.5 × 19.25 mm	83.5°
	0.6 × 20.67 mm	82.3°

Imbibition experiment of liquid-displacing-air in a single channel: Single-channel models with different diameters in the original water-wet state are selected. After water (or oil) is filtered and dropped on one side of the single channel, the change of water front within the channel is monitored by a Leica M165FC stereo fluorescence microscope at the frame rate of 18 frames per second. The imbibition distance is then plotted versus the time, from which the change of imbibition rate is accordingly observed. When water displaces air, the diameter of the single channel remains constant, while the length is changed for comparison; when oil displaces air, the length of the single channel remains unchanged, but the diameter of the channel is changed for comparison.

Imbibition experiment of water-displacing-oil in a single channel: Single channels with different tube diameters and lengths are selected for imbibition experiments of water-displacing-oil. Under the initial condition of saturated kerosene, the 0.05 mL filtered DI-water droplet is dropped on one side of a single channel, and curves of the imbibition distance as a function of time are obtained by the method above. By comparing experimental results and theoretical models, it is investigated the effects of tube diameter, interfacial tension, and viscosity on the imbibition process.

3. Results and Discussion

3.1. Liquid-Displacing-Air

In imbibition studies, both the Washburn model for imbibition of liquid-displacing-gas and the Poiseuille model for imbibition of liquid-displacing-liquid need to meet the laminar flow condition, and the following steps are intended to determine the flow state of this study.

The determination expression of laminar flow is shown as Equation (5) [26]:

$$R_e = \frac{\rho v d}{\mu} \tag{5}$$

Therein, R_e represents Reynolds number; ρ denotes liquid density; v represents average velocity; d is diameter of circular single channel; μ indicates the liquid viscosity.

The Equation (5) shows that Reynolds number is relative to ρ, d, μ, and v, and the v is affected by other factors shown as Equation (6) which is derived from Equation (1).

$$v = \sqrt{\frac{R\sigma \cdot cos\theta}{8\mu}} \cdot t^{-0.5} \tag{6}$$

According to the critical Reynolds number of laminar flow, the condition satisfied the laminar flow is $R_e < 2300$. According to Equations (5) and (6), the Reynolds number increases with the increase of density, diameter and interfacial tension, as well as the decrease of viscosity and contact angle. Therefore, the Reynolds number is most likely to break through critical Reynolds number in the imbibition process of water-displacing-air with large-diameter tube. If imbibition experiments of water-displacing-air meet the laminar flow condition, so do that of oil-displacing-air and water-displacing-oil.

It is assumed that imbibition of water-displacing-air occurs in a 0.5 mm single channel, and the Reynolds number is exactly 2300, then the calculated limit imbibition rate is 4.6 m/s. According to Equation (6), the imbibition rate decreases with time, and the time interval when the initial imbibition rate is greater than the limit one is:

$$t < 1.38 \times 10^{-5}\ \text{s}$$

This time interval is short enough to identify the flow as laminar.

Based on the analysis above, the imbibition process of liquid-displacing-air is in line with the Washburn model's assumptions so that the Washburn model can be used for analysis of the process, and Poiseuille model may be accordingly utilized for prediction on the imbibition process of water-displacing-oil.

In imbibition experiments of liquid-displacing-air, there is no other force except for capillary force. Moreover, the influence of gravity on a liquid droplet is also insignificant compared with the capillary force. Therefore, the displacement force is considered to come only from capillary force.

The experimental results obtained through the microfluidic platform equipped with a fluorescent stereo microscope (Leica M165 FC) and the corresponding prediction curves obtained from Equation (1) are shown in Figure 2, and results of the imbibition distance as a function of $\sqrt{t}$ are shown in Figure 3.

Figures 2 and 3 show that curves of the imbibition distance as a function of time in imbibition experiments liquid-displacing-air with circular single channels fit well with the Washburn model. In the imbibition process that water (or oil) droplet enters the single channel initially saturated with air, the imbibition rate, which varies with the diameter of single channel, decreases gradually with the increase of imbibition distance, and the imbibition distance is linear with the square root of time. Under the condition that the diameters of single channels are all 0.3 mm, four curves of water-displacing-air fluctuate within an acceptable range, indicating that the length of single channel has no influence on imbibition behavior of liquid-displacing-air to some extent.

According to the imbibition distance as a function of time, the imbibition rate can be determined as shown in Figure 4.

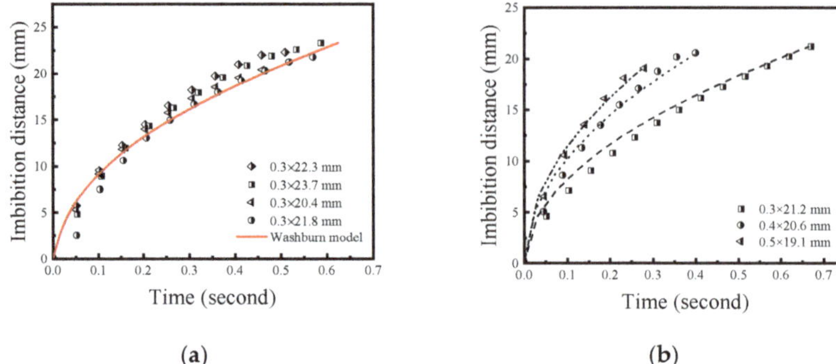

Figure 2. Diagram of the imbibition distance as a function of time in imbibition experiments of liquid-displacing-air in circular single channels. ((**a**) shows the experimental results of water-displacing-air, (**b**) shows the experimental results of oil-displacing-air; these curves and point sets represent Washburn model's prediction curves and the experimental results, respectively).

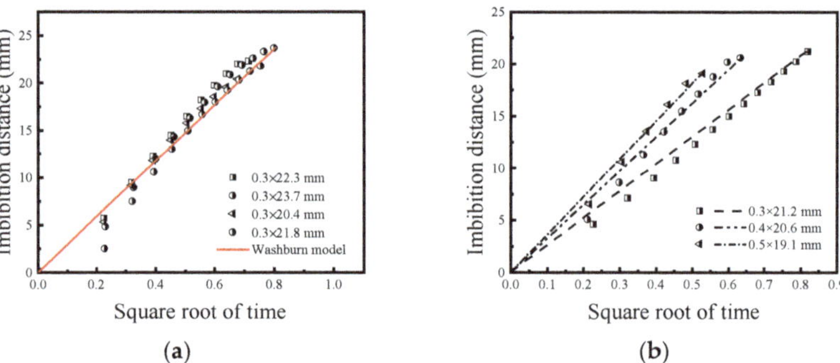

Figure 3. Diagram of the imbibition distance as a function of $\sqrt{t}$ in imbibition experiments of liquid-displacing-air in circular single channels. ((**a**) shows the experimental results of water-displacing-air, (**b**) shows the experimental results of oil-displacing-air; these curves and point sets represent Washburn model's prediction curves and the experimental results respectively).

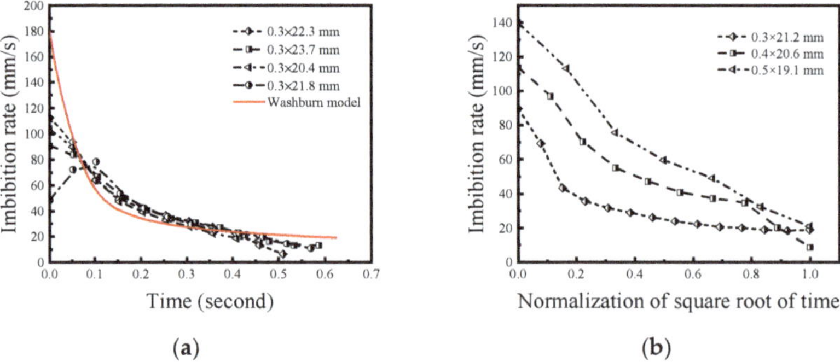

Figure 4. Diagram of the imbibition rate as a function of time. ((**a**) shows the experimental results of water-displacing-air, (**b**) shows the experimental results of oil-displacing-air; these curves and point sets represent Washburn model's prediction curves and the experimental results respectively).

Figure 4a shows that imbibition rates obtained from experiments agree well with the Washburn model after 0.1 s. Mismatch within 0.1 s is likely due to the inertial effect when water (or oil) enters the channel [27]. From the derivation of Equation (1), the capillary force remains unchanged during the whole imbibition process, so the liquid initially imbibed into the single channel produces a great acceleration under the effect of constant capillary force. However, in the actual imbibition process, the velocity of the front in the single channel is affected by the inertance of the DI-water remaining at the inlet.

Figure 4b shows that the tendencies of the imbibition rate, which decrease with time, are the same in different single channels with different diameters in imbibition experiment of oil-displacing-air. In addition, the imbibition rate increases for increased diameters of the single channel, and the increasing rate of imbibition rate reduces with the diameter of the single channel which reflecting that the imbibition rate does not increase infinitely with the increase of tube diameter. The Washburn model predicts such behavior. The good match of Washburn model and experimental results during imbibition of liquid-displacing-air indicates that the wettability of the channel wall remains unchanged during the imbibition.

3.2. *Water-Displacing-Oil*

Imbibition of a water-displacing-oil is different from that of liquid-displacing-air. As for the former one, the viscosity of the displaced phase can't be neglected. In fact, the model should be modified by introducing the viscosity of the displaced phase, shown in Equation (4). In the imbibition experiments of water-displacing-oil, the focus is to determine how the length and diameter of single channel influence the imbibition behavior, so viscosity of liquid and wettability of channel wall should keep constant. As shown in imbibition experiments of liquid-displacing-air in Table 1 and Figure 3, wettability of the inner wall remains unchanged during imbibition, so is the viscosity of liquid.

The experimental results of the imbibition distance as a function of time in imbibition experiments of water-displacing-oil are obtained through the microfluidic platform, as shown in Figure 5.

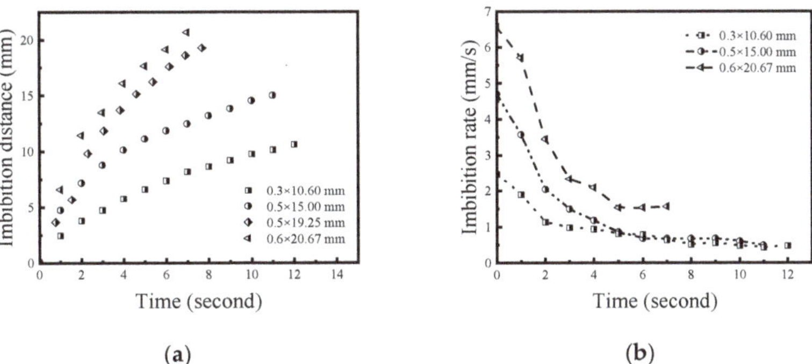

Figure 5. (**a**) Diagram of the imbibition distance as a function of time in water-displacing-oil experiments in circular single channels. (**b**) Diagram of the imbibition rate as a function of time. (Three of the four sets of experiment data specified in (**a**) are selected randomly for the subsequent fitting analysis, and the remaining one set of data thereof is for verification on fitting steps, as indicated in (**b**)).

Figure 5 shows that the imbibition rate is in the same tendency in several groups of imbibition experiments in which the lower-viscosity DI-water displaces the higher-viscosity kerosene, that is, the imbibition rate gradually decreases during spontaneous imbibition, and increases with the increase of diameter of single channel, of which the effect on imbibition rate is widely known. Figure 5a also shows that in the single channel with the same diameter, the length has a great influence on imbibition rate, which is caused by the viscosity of displaced phase.

Combined with the parameters in Table 1 and Equation (4), the prediction curve of Poiseuille model in a single channel with diameter of 0.3 mm is drawn, and the result of imbibition experiment and prediction curve are compared, as shown in Figure 6.

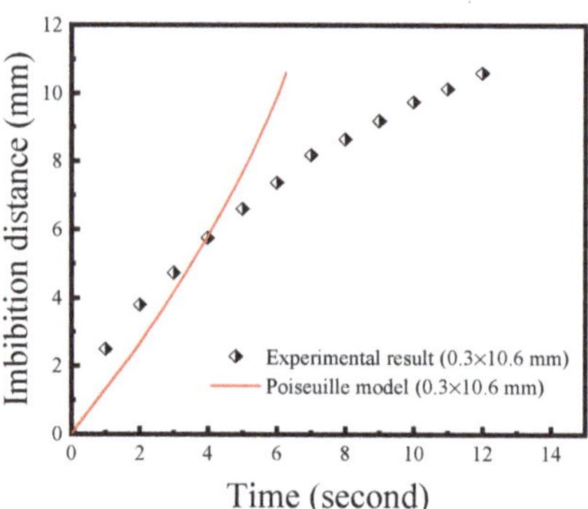

Figure 6. Comparison diagram of experimental data and Poiseuille model curve of imbibition experiments of water-displacing-oil in circular single channel.

Figure 6 shows that curve of the imbibition distance as a function of time obtained by Poiseuille model is different from that of experimental result, which was also observed in Yang's paper [23].

The reasons of this abnormal phenomenon are as follows. In the Poiseuille model it is assumed that displacement force only comes from capillary force at water-oil interface, displacement resistance comes from viscous force generated by flow of oil and water, and other forces are ignored. In the actual core displacement experiment, displacement pressure difference that is far greater than capillary force is applied to the core so that the influence of other displacement resistances on flow can be ignored. Moreover, the imbibition experiments with single-channel model are subject to the trend of imbibition rate in the gradual decrease both in case that the lower-viscosity liquid displaces the higher-viscosity one and the higher-viscosity liquid displaces the lower-viscosity one. While the Equation (4) indicates that such a decreasing trend for imbibition rate is normal when the higher-viscosity liquid displaces the lower-viscosity one. However, the decreasing trend of imbibition rate is abnormal in actual imbibition experiments of the lower-viscosity liquid displacing the higher-viscosity one compared to the increasing trend shown in Equation (4). Therefore, Equation (4) needs to be modified on combination of the actual experimental conditions and specific forces.

In the imbibition experiment of water-displacing-oil, a single channel initially saturated with kerosene is placed on a horizontal glass base, and water droplet is quantitatively dropped at one end of the single channel (inlet end). Water is imbibed into the single channel due to the capillary farce. At the same time, kerosene is displaced from the other end of the single channel (outlet end), and imbibed reversely to the inlet end of the single channel along the angle between the single channel and the horizontal glass base. The reverse imbibition at the outer angle is one kind of imbibition of liquid-displacing-air. Thus, capillary force, viscous force and the force caused by the reverse imbibition are the three main forces.

The mechanism of the third force is as follows. Due to reverse imbibition of kerosene a small concave meniscus will appear at the outlet end of the single channel, and the curvature of concave meniscus decreases with the increase of the volume of kerosene gathered at the angle outside the single channel. During the period of concave meniscus, the force caused by the reverse imbibition stimulates the imbibition of water-displacing-oil inside the single channel. Then, the concave meniscus disappears and transition to a convex meniscus occurs until the kerosene accumulated at the angle reaches the limit volume allowed by wettability. At this moment, the force produced by the reverse imbibition hinders the internal imbibition Therefore, in the imbibition experiment conducted in capillary tubes, displaced phase not only produces a viscous force inside the capillary tube, but also generates an additional force when accumulating outside the tube.

The reverse imbibition is affected by the wettability both of the single channels and the glass base that is one component of the microfluidic platform, all of which determines the liquid storage capacity of the angle outside single channel. The liquid storage capacity is related to the limit volume of concave meniscus. In this study, wettability of glass base and single channels keeps constant, for the glass base is cleaned and dried once an imbibition experiment is completed with it, and every new single channel used for each experiment is provided by the same supplier.

After analyzing the forces, Poiseuille model can be modified by introducing the third force. Therefore, the differential pressure ΔP is modified shown as Equation (7):

$$\Delta P = P_c + P_c' \tag{7}$$

P_c' represents the third force, which is related to imbibition distance.

After introducing the third force, the key is to determine its expression.

According to the specific experimental situation, the following hypothesis is made: the kerosene displaced from the single channel instantly fills the external angle. This is since the imbibition at the angle is a kind of imbibition of liquid-displacing-air, of which the imbibition rate is much higher than that of water-displacing-oil.

From the geometric analysis, it can be known that the shapes of cross section of kerosene gathered at the included angle outside the single channels are similar when the normalized imbibition distances in single channels with different diameters are the same. Moreover, the shape of the cross section is only related to the normalized imbibition distance on premise of the same wettability.

Thus, P_c' can be expressed as follows in Equation (8):

$$P_c' = a \times \left(\frac{L}{L_{channel}}\right)^n + b \tag{8}$$

Therein, a, b and n represent 3 characteristic parameters.

Then, the ultimate expression is shown in Equation (9):

$$\frac{R^2}{8}\left[P_c + a \times \left(\frac{L}{L_{channel}}\right)^n + b\right] \times t = \frac{1}{2}(\mu_w - \mu_{nw})L^2 + \mu_{nw} L_{channel} L \tag{9}$$

According to the analysis in the Appendix A, the value of the characteristic parameter n is 0.5. And then, characteristic parameter a and b need to be determined by following steps.

When the imbibition process is finished, P_c' can be expressed as Equation (10):

$$P_{cend}' = a + b \tag{10}$$

Therein, P_{cend}' indicates the P_c' at the end of imbibition.

According to Equation (9), all parameters other than characteristic parameter a and b are known during the whole imbibition process, so P_{cend}' can be calculated when the imbibition process is finished. P_{cend}' in one experiment is constant, so is the sum of a plus b. Therefore, the optimal fitting curve is obtained by constantly adjusting the values of a and b to fit the experimental data. Finally, characteristic parameters a and b are determined.

Thus, characteristic parameter n, a and b are determined in each imbibition experiment, as shown in Table 2.

Table 2. Values of characteristic parameters a and b.

	0.3 × 10.60 mm	0.5 × 15.00 mm	0.6 × 20.67 mm
a	−18.50	−21.07	−34.08
b	13.63	14.19	24.12

According to these parameters, the modified curve of Poiseuille model is drawn in Figure 7.

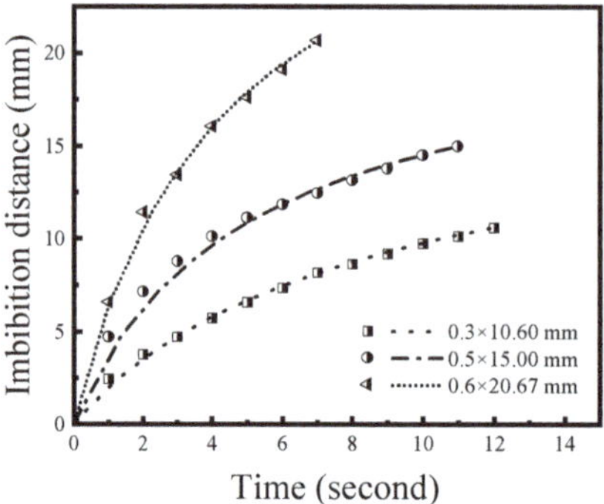

Figure 7. Comparison diagram between the fitted model curves and the corresponding experimental results (These curves represent the modified Poiseuille model curves, and the series of points represent the corresponding experimental data).

Figure 7 shows that the modified Poiseuille model can better reflect the law of imbibition rate in the imbibition process that the lower-viscosity liquid displaces the higher-viscosity liquid.

In the three groups of imbibition experiments of water-displacing-oil above, the prediction curve is adjusted and fitted with the experimental data by changing characteristic parameter a and b, then the best prediction curve is obtained, and the corresponding characteristic parameter a and b are concluded. As a result, all parameters in Equation (9) are known. Furthermore, a constant dimensionless parameter P'_{c0}/P_c (details refer to Table 3), which is consistent with that in other imbibition experiments of water-displacing-oil, is found and used for calculating characteristic parameters a and b.

Table 3. Parameters at the initial and terminal positions of imbibition.

	0.3 × 10.60 mm	**0.5 × 15.00 mm**	**0.6 × 20.67 mm**
P'_{c0} (or b)	13.63	14.19	24.12
P'_{c0}/P_c	1.34	1.28	1.29

(P'_{c0}/P_c represents the ratio of the third force at the angle outside the single channel to the capillary force inside the single channel when $t = 0$).

In other imbibition experiments under same conditions, characteristic parameter a and b are determined according to the constant dimensionless parameter.

Specific steps are as follows.

When $t = 0$, Equation (8) can be expressed as Equation (11):

$$P'_{c0} = b \tag{11}$$

P'_{c0} and the ratio of P'_{c0} to P_c are calculated and shown in Table 3:

Table 3 shows that the average value of P'_{c0}/P_c is 1.30. Capillary force can be easily calculated, and P'_{c0} can be calculated as well according to the value of P'_{c0}/P_c. Therefore, characteristic parameter b is obtained. In addition, P'_{cend} equal to the sum of a and b is calculated according to Equation (9). Thus, characteristic parameter a is calculated. Upon conclusion of the step above, the Equation (9) utilized in the three imbibition experiments is, respectively, determined.

To verify the practicability and accuracy of this modified Poiseuille model, characteristic parameter a and b in other imbibition experiments of water-displacing-oil are calculated through the above method, and the parameters used in this verification case include the dimensionless parameter shown in Table 3 and the measured parameters shown in Table 1 and Figure 5. The prediction curve of modified Poiseuille model is obtained, as shown in Figure 8.

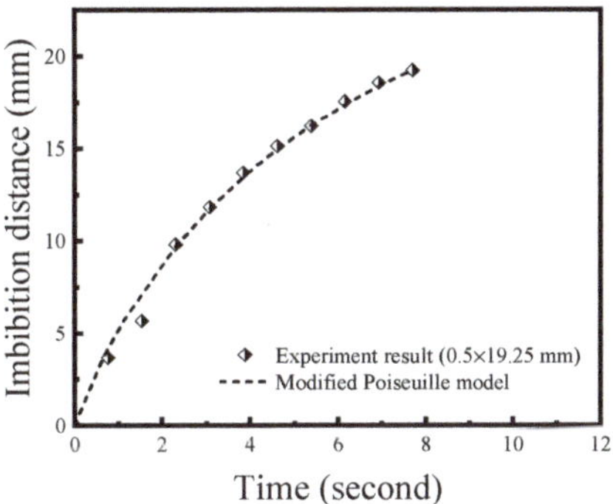

Figure 8. Comparison diagram between the prediction curve of modified Poiseuille model and experimental result in one 0.5 × 19.25 mm single channel. These experiment data are derived from Figure 5a.

Figure 8 shows that the modified Poiseuille model has a good prediction effect. According to the above analysis of imbibition of water-displacing-oil in a single channel, when the viscosity of the displaced phase cannot be ignored, the accumulation effect by the external liquid on imbibition process cannot be ignored as well.

In the modified Poiseuille model, the effects of the length and diameter of the single channel and additional force come from the reverse imbibition on the imbibition are considered. After introducing the additional force called as the third force above, the modified Poiseuille model can describe the imbibition process of water-displacing-oil, which is used to predict the imbibition process where the lower-viscosity liquid displaces the higher-viscosity liquid. The accumulation effect on imbibition in the process that the higher-viscosity liquid displaces the lower-viscosity liquid is likely to exist, and it is ignored for the tendency of imbibition rate is unchanged (although the accumulation effect is considered).

4. Conclusions

In this study, spontaneous imbibition of liquid-displacing-air and water-displacing-oil is respectively observed through the microfluidics. According to the study on the reverse imbibition occurred from the displaced phase at the angle outside single channel, Poiseuille model describing the liquid-liquid spontaneous imbibition is modified.

(1) In experiments of spontaneous imbibition of liquid-displacing-air, the results are in line with Washburn model and match well with previous results using the vapor whose viscosity is significantly lower than the liquid.

(2) The experimental results in the lower-viscosity liquid (DI-water) displacing the higher-viscosity one (kerosene/oil) show that the imbibition rate decreases with time, which is not in line with the increasing trend of imbibition rate indicated in Equation (4).

(3) On the basis of the Poiseuille model, a new model is proposed that includes the additional force from the accumulated fluid flowing out of the single channel. The modified Poiseuille model matches well with the experimental results and shows that a good prediction effect in the water-displacing-oil imbibition experiment.

Author Contributions: Conceptualization, W.D. and T.L.; Data curation, W.D.; Formal analysis, W.D.; Funding acquisition, T.L.; Methodology, W.D.; Supervision, T.L., F.Z. and J.L.; Validation, S.Y.; Writing—original draft, W.D.; Writing—review & editing, T.L. All authors have read and agreed to the published version of the manuscript.

Funding: This research was funded by the Strategic Cooperation Technology Projects of CNPC and CUPB (ZLZX2020-01).

Institutional Review Board Statement: Not applicable.

Informed Consent Statement: Not applicable.

Data Availability Statement: Not applicable.

Acknowledgments: This work was financially supported by the Strategic Cooperation Technology Projects of CNPC and CUPB (ZLZX2020-01).

Conflicts of Interest: The authors declare no conflict of interest.

Appendix A

Characteristic Parameter n

When P_c' is positive, it is obvious that the volume of kerosene accumulated at the angle outside the single channel has not reached the limit of aggregation volume allowed by wettability ($P_c' = 0$), and the displaced kerosene at the angle can also conduct spontaneous imbibition along the outer tube wall. It is assumed to set the capillary force providing spontaneous imbibition at the external angle to P_c''. The larger P_c'' is, the faster the rate of reverse imbibition of kerosene is, the greater the force on internal imbibition of water-displacing-oil is, and the higher the imbibition rate of internal imbibition is. Thus, P_c' and P_c'' change in the same way. Therefore, P_c'' is analyzed to determine the characteristic parameter n that reflects the trend of P_c'.

According to experimental conditions, the shape of cross section of kerosene outside the single channel is shown in Figure A1.

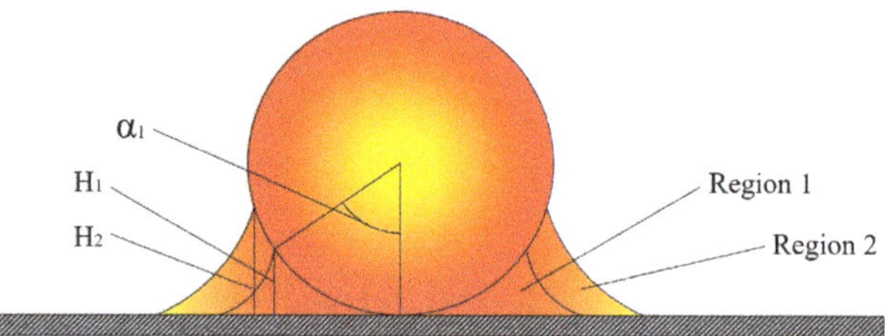

Figure A1. Schematic diagram of the cross section.

Due to the difference between imbibition of liquid-displacing-air and that of water-displacing-oil, the displacement resistance of kerosene-displacing-air is far less than that of water-displacing-oil, so that the outer kerosene is evenly distributed at the angle which named region 1 in a very short time. And then, kerosene is displaced out of the single channel, and reversely imbibed along the angle at region 2. The area of region 1 is equal to the area of region 2. Assuming the imbibition distance inside the single channel is L, the cross-sectional area at the angle is shown in Equation (A1):

$$S = \frac{\pi R^2 \cdot L}{L_{channel}} \tag{A1}$$

As shown in this figure, the height of the cross section of kerosene at the angle corresponding to a certain imbibition distance is H, and the corresponding angle is α. The contact angle between the external angle and the horizontal base is θ'. Therefore, the cross-sectional area at the angle is also expressed as Equation (A2):

$$S = 2 \times \left[\frac{H^2}{2tan\theta'} + \frac{1}{2}(H+R)\cdot R\cdot sin\alpha - \frac{1}{2}\cdot\frac{\alpha\cdot\pi}{180}\cdot R^2 \right] \quad \text{(A2)}$$

The relationship between the height of cross section and the imbibition distance can be obtained by substituting values, as shown in Figure A2.

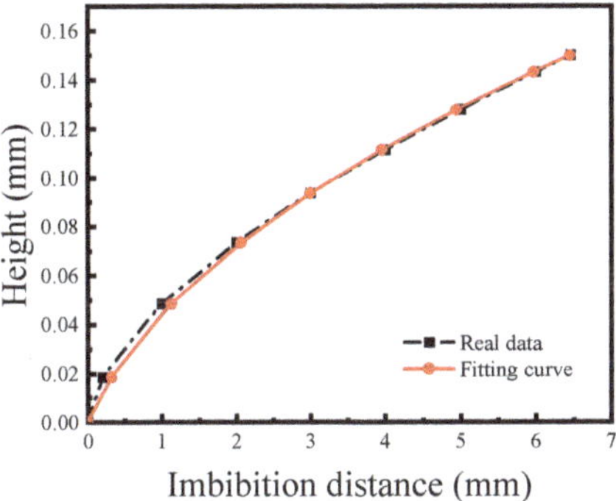

Figure A2. The curves of the height of cross section as a function of imbibition distance (the black curve is obtained according to Equations (A1) and (A2), the red one is a fitting curve).

The expression of the fitting curve is shown in Equation (A3):

$$L = 195.0718 \times H^2 + 13.7926 \times H \quad \text{(A3)}$$

This expression reflects the relationship between the imbibition distance in the single channel and the height of the cross-sectional area at the angle outside the single channel.

And then, the relationship between P_c'' and imbibition distance is obtained by analyzing the relationship between P_c'' and H.

Combined with the analysis above, the imbibition rate of water-displacing-oil inside the single channel is far less than that of kerosene-displacing-air at the angle outside the single channel. Therefore, the following assumption is made: when the initial length of water phase imbibed into the single channel is L_0, the kerosene displaced from the outlet end must be imbibed in reverse at a very fast speed and evenly distributed at the external angle (region 1). At this moment, the height of the cross-section at the external angle is H_1. When the water phase is imbibed the second length of L_0, the displaced kerosene will be reversely imbibed at region 2. Therefore, the expression of P_c'' is given as Equation (A4):

$$\begin{cases} P_{c1}'' = \dfrac{2\sigma R\cdot\frac{\alpha_1\cdot\pi}{180}\cdot cos\theta'' + 2\sigma\cdot\left(\frac{H_1}{tan\theta'} + Rsin\alpha_1\right)\cdot cos\theta'}{\frac{\pi R^2 L_0}{L_{channel}}} & (k=1) \\ P_{ck}'' = \dfrac{2\sigma R\cdot(\alpha_k-\alpha_{k-1})\cdot\frac{\pi}{180}\cdot cos\theta'' + 2\sigma\cdot\left[\frac{H_k-H_{k-1}}{tan\theta'} + R(sin\alpha_k - sin\alpha_{k-1})\right]\cdot cos\theta'}{\frac{\pi R^2 L_0}{L_{channel}}} & (k>1) \end{cases} \quad \text{(A4)}$$

Therein, k represents the ratio of imbibition distance to L_0; α_k and r_k, respectively, represent the height and angle when the corresponding imbibition distance is $k\cdot L_0$, as shown in Figure A1. θ' represents the contact angle between kerosene and the glass base; θ'' represents the contact angle between kerosene and the outer wall of the single channel.

The height H_k in Equation (A4) can be expressed by imbibition distance L, as shown in Equation (A3). Thus, P_c'' in Equation (A4) is a function of L. The relationship between P_c'' and the imbibition distance (L) is shown in Figure A3.

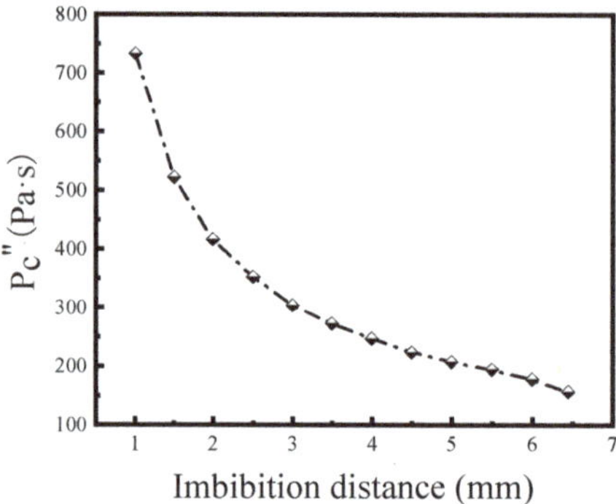

Figure A3. The curve of P_c'' as a function of imbibition distance.

The figure above shows that P_c'' has a linear relationship with the square root of imbibition distance (L), and P_c'' decreases with the increase of imbibition distance, so the characteristic parameter a in Equation (8) is negative. According to the similarity between P_c' and P_c'', the characteristic parameter n in Equation (8) is determined to be 0.5.

Expression (8) and Expression (9) become the following expression as shown in Equations (A5) and (A6):

$$P_c' = a \times \left(\frac{L}{L_{channel}}\right)^{0.5} + b \tag{A5}$$

$$\frac{R^2}{8} \cdot \left[P_c + a \times \left(\frac{L}{L_{channel}}\right)^{0.5} + b\right] \cdot t = \frac{1}{2} \cdot (\mu_w - \mu_{nw}) \cdot L^2 + \mu_{nw} L_{channel} L \tag{A6}$$

Equation (A6) is a modified model considering the third force at the angle outside the single channel. The values of characteristic parameters a and b can be obtained by the curves of time and distance in the imbibition process, and then a modified Poiseuille model for imbibition of water-displacing-oil can be obtained, which can be used to predict the imbibition process under the same conditions.

References

1. Liang, T.; Zhao, X.; Yuan, S.; Zhu, J.; Liang, X.; Li, X.; Zhou, F. Surfactant-EOR in tight oil reservoirs: Current status and a systematic surfactant screening method with field experiments. *J. Pet. Sci. Eng.* **2021**, *196*, 108097. [CrossRef]
2. Sun, L.; Zou, C.; Jia, A.; Wei, Y.; Zhu, R.; Wu, S.; Guo, Z. Development characteristics and orientation of tight oil and gas in China. *Pet. Explor. Dev.* **2019**, *46*, 1073–1087. [CrossRef]
3. Chen, W.; Zhang, Z.; Liu, Q.; Chen, X.; Opoku Appau, P.; Wang, F. Experimental Investigation of Oil Recovery from Tight Sandstone Oil Reservoirs by Pressure Depletion. *Energies* **2018**, *11*, 2667. [CrossRef]
4. Sheng, J.J. What type of surfactants should be used to enhance spontaneous imbibition in shale and tight reservoirs? *J. Pet. Sci. Eng.* **2017**, *159*, 635–643. [CrossRef]
5. Zou, C.; Zhang, G.; Yang, Z.; Tao, S.; Hou, L.; Zhu, R.; Yuan, X.; Ran, Q.; Li, D.; Wang, Z. Concepts, characteristics, potential and technology of unconventional hydrocarbons: On unconventional petroleum geology. *Pet. Explor. Dev.* **2013**, *40*, 413–428. [CrossRef]
6. Mason, G.; Morrow, N.R. Developments in spontaneous imbibition and possibilities for future work. *J. Pet. Sci. Eng.* **2013**, *110*, 268–293. [CrossRef]
7. Lucas, R. Ueber das Zeitgesetz des kapillaren Aufstiegs von Flüssigkeiten. *Kolloid-Zeitschrift* **1918**, *23*, 15–22. [CrossRef]
8. Washburn, E.W. The Dynamics of Capillary Flow. *Phys. Rev.* **1921**, *17*, 273. Available online: https://journals.aps.org/pr/abstract/10.1103/PhysRev.17.273 (accessed on 9 November 2021). [CrossRef]
9. Zhang, J.; Sun, Z.; Wang, X.; Kang, X. Study on the Oil Displacement Effect and Application of Soft Microgel Flooding Technology. In Proceedings of the SPE Middle East Oil & Gas Show and Conference, event canceled, 28 November–1 December 2021.
10. Qin, T.; Fenter, P.; AlOtaibi, M.; Ayirala, S.; AlYousef, A. Pore-Scale Oil Connectivity and Displacement by Controlled-Ionic-Composition Water-flooding Using Synchrotron X-ray Microtomography. *SPE J.* **2021**, *26*, 3694–3701. [CrossRef]
11. Mirzaei-Paiaman, A.; Masihi, M. Scaling Equations for Oil/Gas Recovery from Fractured Porous Media by Counter-Current Spontaneous Imbibition: From Development to Application. *Energy Fuels* **2013**, *27*, 4662–4676. [CrossRef]

12. Wang, F.; Zhao, J. A mathematical model for co-current spontaneous water imbibition into oil-saturated tight sandstone: Up-scaling from pore-scale to core-scale with fractal approach. *J. Pet. Sci. Eng.* **2019**, *178*, 376–388. [CrossRef]
13. Li, C.; Shen, Y.; Ge, H.; Zhang, Y.; Liu, T. Spontaneous imbibition in fractal tortuous micro-nano pores considering dynamic contact angle and slip effect: Phase portrait analysis and analytical solutions. *Sci. Rep.* **2018**, *8*, 3919. [CrossRef] [PubMed]
14. Hilpert, M. Effects of dynamic contact angle on liquid infiltration into horizontal capillary tubes: (Semi)-analytical solutions. *J. Colloid Interface Sci.* **2009**, *337*, 131–137. [CrossRef] [PubMed]
15. Hilpert, M. Effects of dynamic contact angle on liquid infiltration into inclined capillary tubes: (Semi)-analytical solutions. *J. Colloid Interface Sci.* **2009**, *337*, 138–144. [CrossRef]
16. Soares, E.J.; Carvalho, M.S.; Mendes, P.R.S. Immiscible Liquid-Liquid Displacement in Capillary Tubes. *J. Fluids Eng.* **2005**, *127*, 24–31. [CrossRef]
17. Yang, W.; Fu, C.; Du, Y.; Xu, K.; Balhoff, M.T.; Weston, J.; Lu, J. Dynamic Contact Angle Reformulates Pore-Scale Fluid-Fluid Displacement at Ultralow Interfacial Tension. *SPE J.* **2021**, *26*, 1278–1289. [CrossRef]
18. Tian, W.; Wu, K.; Chen, Z.; Lai, L.; Gao, Y.; Li, J. Effect of Dynamic Contact Angle on Spontaneous Capillary-Liquid-Liquid Imbibition by Molecular Kinetic Theory. *SPE J.* **2021**, *26*, 2324–2339. [CrossRef]
19. Sugar, A.; Torrealba, V.; Buttner, U.; Hoteit, H. Assessment of Polymer-Induced Clogging Using Microfluidics. *SPE J.* **2021**, *26*, 3793–3804. [CrossRef]
20. Xu, K.; Liang, T.; Zhu, P.; Qi, P.; Lu, J.; Huh, C.; Balhoff, M. A 2.5-D glass micromodel for investigation of multi-phase flow in porous media. Lab on a Chip. *R. Soc. Chem.* **2017**, *17*, 640–646.
21. Han, A.; Mondin, G.; Hegelbach, N.G.; de Rooij, N.F.; Staufer, U. Filling kinetics of liquids in nanochannels as narrow as 27 nm by capillary force. *J. Colloid Interface Sci.* **2006**, *293*, 151–157. [CrossRef]
22. Yang, L.-J.; Yao, T.-J.; Tai, Y.-C. The marching velocity of the capillary meniscus in a microchannel. *J. Micromech. Microeng.* **2004**, *14*, 220–225. [CrossRef]
23. Yang, D.; Krasowska, M.; Priest, C.; Ralston, J. Dynamics of capillary-driven liquid–liquid displacement in open microchannels. *Phys. Chem. Chem. Phys.* **2014**, *16*, 24473–24478. [CrossRef] [PubMed]
24. Yang, D.; Krasowska, M.; Priest, C.; Popescu, M.N.; Ralston, J. Dynamics of Capillary-Driven Flow in Open Microchannels. *J. Phys. Chem. C* **2011**, *115*, 18761–18769. [CrossRef]
25. Chen, T. Capillary force-driven fluid flow of a wetting liquid in open grooves with different sizes. In *Fourteenth Intersociety Conference on Thermal and Thermomechanical Phenomena in Electronic Systems (ITherm)*; IEEE: Orlando, FL, USA, 2014; pp. 388–396.
26. Deng, Y.; Li, J.; Meng, J.; Cao, Y.; Zhao, L.; Zhao, J. Study on Marine Pipelines Forced Vibration and Vortex-Induced Vibration in Uniform flow and Combined Flow at Different Reynolds Number. In Proceedings of the 31st International Ocean and Polar Engineering Conference, Rhodes, Greece, 20–25 June 2021.
27. Tian, W.T.; Wu, K.; Chen, Z.; Gao, Y.; Gao, Y.; Li, J. Inertial Effect on Spontaneous Oil-Water Imbibition by Molecular Kinetic Theory. In Proceedings of the SPE Europec Featured at 82nd EAGE Conference and Exhibition, Amsterdam, The Netherlands, 18–21 October 2021; The Society of Petroleum Engineers (SPE): Dallas, TX, USA, 2021.

MDPI

Article

Experimental Study of Temperature Effect on Methane Adsorption Dynamic and Isotherm

Yongchun Zhang [1,2], Aiguo Hu [1,2], Pei Xiong [2], Hao Zhang [1,*] and Zhonghua Liu [3,*]

1 College of Energy Resource, Chengdu University of Technology, Chengdu 610059, China; hbjzyc@126.com (Y.Z.); huaiguo19842004@163.com (A.H.)
2 Research Institute of Engineering and Technique, Sinopec Huabei Sub–Company, Zhengzhou 450006, China; xiongpei_2004@126.com
3 School of Petroleum and Natural Gas Engineering, Chongqing University of Science and Technology, Chongqing 401331, China
* Correspondence: zhanghao@cdut.edu.cn (H.Z.); liuzhonghua@cqust.edu.cn (Z.L.)

Abstract: Knowing the methane adsorption dynamic is of great importance for evaluating shale gas reserves and predicting gas well production. Many experiments have been carried out to explore the influence of many aspects on the adsorption dynamic of methane on shale rock. However, the temperature effect on the adsorption dynamic as a potential enhanced shale gas recovery has not been well addressed in the publications. To explore the temperature effect on the adsorption dynamic of methane on gas shale rock, we conducted experimental measurement by using the volumetric method. We characterized the adsorption dynamic of methane on gas shale powders and found that the curves of pressure response at different pressure steps and temperatures all have the same tendency to decrease fast at first, then slowly in the middle and remain stable at last, indicating the methane molecules are mainly adsorbed in the initial stage. Methane adsorption dynamic and isotherm can be well fitted by the Bangham model and the Freundlich model, respectively. The constant z of the Bangham model first decreases and then increases with equilibrium pressure increasing at each temperature, and it decreases with temperature increasing at the same pressure. The adsorption rate, constant k of the Bangham model, is linearly positively correlated with the natural log of the equilibrium pressure, and it decreases with temperature increasing at the same pressure. Constant K and *n* of the Freundlich model all decrease with temperature increasing, indicating that low temperatures are favorable for methane adsorption on shale powders, and high temperatures can obviously reduce constant K and *n* of the Freundlich model. Finally, we calculated isosteric enthalpy and found that isosteric enthalpy is linearly positively correlated with the adsorption amount. These results will be profoundly meaningful for understanding the mechanism of methane adsorption dynamic on shale powders and provide a potential pathway to enhance shale gas recovery.

Keywords: adsorption dynamic; Longmaxi shale; Bangham model; Freundlich model; adsorption isotherm; isostatic enthalpy

Citation: Zhang, Y.; Hu, A.; Xiong, P.; Zhang, H.; Liu, Z. Experimental Study of Temperature Effect on Methane Adsorption Dynamic and Isotherm. *Energies* **2022**, *15*, 5047. https://doi.org/10.3390/en15145047

Academic Editor: Daoyi Zhu

Received: 6 June 2022
Accepted: 4 July 2022
Published: 11 July 2022

1. Introduction

Shale gas, as one of the most promising unconventional natural gas resources, has attracted increasing attention in recent years. As we know, adsorbed gas is a major type of shale gas stored in shale gas reservoir [1,2], which distributes from 20% to 85% of the total gas in shale gas reservoirs [3,4]. Therefore, characterizing the adsorption dynamic and knowing the temperature influence is of great significance for exploring the methane adsorption mechanism on gas shale.

In recent years, the experiments of methane adsorption dynamic have been conducted in many publications by using volumetric and gravimetric methods [5–7]. Gas-in-place evaluation can be accurately calculated based on the experimental measurements from laboratory [8]. Many scholars have pointed out that the methane adsorption on the gas shale is closely correlated with kerogen types [9], TOC [10], mineral constituents of shale rock [11], pore structure [12,13], temperature [14] and moisture [15]. At the same time, the characteristics of methane adsorption and desorption are always revealed by fitting adsorption or desorption equations, such as the Langmuir model [16], Freundlich model [17,18] and modified Dubinin−Radushkevich model [19].

Meanwhile, methane adsorption dynamics have attracted attention from many scholars [1,20–22]. Yuan et al. [23] investigated the pore structure characteristics of Lower Silurian shale and the diffusion behavior of methane molecules, as well as the shale sample size influence on the adsorption and diffusion of methane. Chen et al. [24] studied the mechanisms of methane adsorption in terms of methane adsorption thermodynamics under the condition of high pressure. Rani et al. [25] characterized the adsorption dynamic of methane according to the unipore model and the modified unipore model, respectively. Dasani et al. [8] pointed out the adsorption dynamic of methane mixed with ethane in gas shale samples. Yang et al. [26] published a dynamic adsorption–diffusion equation after dynamic adsorption measurements under the certain condition of a constant pressure and then compared with the instantaneous adsorption–diffusion equation and diffusion equations. However, the adsorption dynamic of methane on Longmaxi gas shale is not still well characterized, and the temperature influence on the methane adsorption dynamic is also not clear.

In this study, for better understanding the temperature effect on methane adsorption dynamic, we conducted the experimental measurements of adsorption dynamic by using the volumetric method at different temperatures, as well as characterizing the adsorption dynamic for methane on gas shale powders and fitting the experimental data by using the Bangham model and the Freundlich model. Finally, we explored the temperature effect on the adsorption dynamic equation, isotherm equation and adsorption thermodynamics. These results are profoundly meaningful for advancing the mechanism of shale gas adsorption dynamic in shale reservoirs.

2. Experimental Design

2.1. Shale Samples

The shale samples from the lower of Silurian Longmaxi formation in northeast of Chongqing were crushed and then went through sieves with diameters of 0.25 and 0.125 mm in sequence. The shale powders between 0.25 and 0.125 mm in diameter were dried in an oven at a constant temperature of 60 °C, but some water may still have been trapped in the pores of shale samples. An amount of 130 g of gas shale powders was prepared for methane adsorption dynamic test. The density of shale samples was 2.56 g/cm^3. Based on the experimental results of low-pressure nitrogen adsorption/desorption, the specific surface area was 26.58 m^2/g, and the pore sizes of the shale sample were mainly distributed in the broad mesoporous region. According to X-ray diffraction, the shale samples mainly contained quartz, clay mineral, dolomite and plagioclase, and their contents were 46.2%, 32.3%, 9.9% and 5.1%, respectively.

2.2. Experimental Apparatus

An experimental apparatus is shown in Figure 1 to test the pressure response during the dynamic process of methane adsorption on shale powders, which was designed using the volumetric method and consisted of many parts, such as methane tank, nitrogen tank, gas compressor, vacuum pump, pressure transducers, reference cell, sample cell, water bath, computer and other valves. The potentiometric pressure transducers were used in this work, and their maximum pressure was 40 MPa, and the precision was 0.25% of the full-scale value. The water bath could work from an indoor temperature to 100 °C, and the control precision was achieved to 0.2 °C. Methane in the methane tank was used for the adsorption dynamic test on the shale powders, and nitrogen contained in the nitrogen tank was applied to check the leak of the experimental setup and obtain the void volume of shale powders loaded in the sample cell. To make the experimental temperature stable, a water bath with a temperature controller was added. To read the pressure response during the experimental procedure, pressure transducers were used, and the data were recorded using a computer.

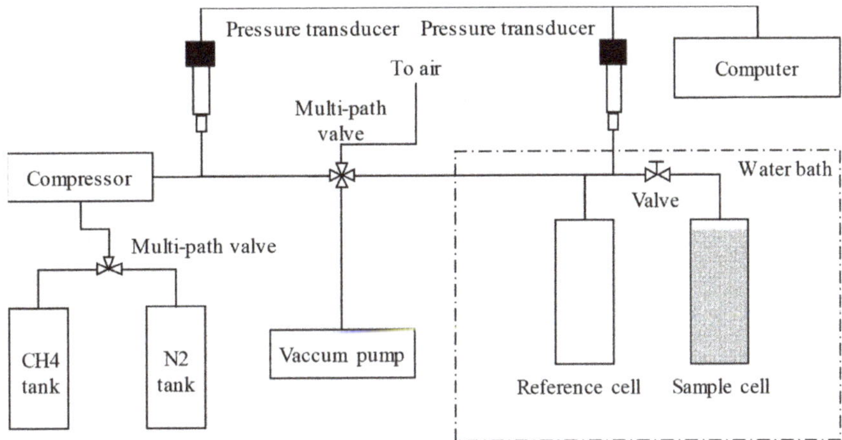

Figure 1. Schematic diagram of experimental apparatus.

2.3. Procedures

To characterize the adsorption dynamic of methane on gas shale, the procedures were shown as follows:

1. Preparation. An amount of 130.00 g of shale powders, whose diameter was between 0.125 and 0.25 mm, was loaded into the sample cell and then heated and maintained at 30, 40 or 50 °C, respectively.
2. Leak check of experimental apparatus. To check the leak of the apparatus at a certain pressure and temperature, nitrogen was used until the pressure reached and maintained stability for two hours, and the temperature remained stable.
3. Free volume determination. Free volume is defined as the space between the shale powders and within the powders and was calculated according to Boyle's law by using the different equilibrium pressures.
4. Apparatus vacuum. The apparatus was vacuumed down to 10^{-5} Pa for 24 h to completely remove the gas molecules from the shale powders.
5. Adsorption dynamic determination. After the pressure of the reference cell full of pure methane remained stable, the reference cell and sample cell were connected by opening the valve between them. Then, the two stable pressures before and after the methane expansion were read and recorded. This procedure should be repeated until an equilibrium pressure of about 30 MPa is reached. The pressure response during the dynamic process of methane adsorption was read and recorded.

3. Mathematical Model

3.1. Dynamic Model

The Bangham model is widely used for modeling methane adsorption dynamic, and it is expressed as the following equation [24,27–29]:

$$q_t = \mathrm{q_e}\left(1 - \mathrm{e}^{-\mathrm{k}t^z}\right) \quad (1)$$

where z is a constant, and k is the adsorption rate of the Bangham equation.

Equation (1) can be shown as the following formula:

$$-\ln\frac{q_e - q_t}{q_e} = kt^z \tag{2}$$

The linear formula of the Bangham equation can be shown as follows:

$$\ln(-\ln\frac{q_e - q_t}{q_e}) = \ln(k) + z\ln(t) \tag{3}$$

From Equation (3), the relationship between $\ln(-\ln\frac{q_e - q_t}{q_e})$ and $\ln(t)$ should be fitted into linear equation, and k and z can be obtained according to the y-intercept of ln(k) and the slope of z.

3.2. Isotherm Model

The Freundlich model is a classical isothermal adsorption model and regarded as an extension of the Henry model [30]. The equation can be expressed as follows:

$$V = KP^n \tag{4}$$

where V is the adsorption capacity per unit mass samples (cc/g), K is the Freundlich constant related to a measure of adsorption capacity (cc/g/MPa), and P is the equilibrium pressure (MPa). n is a constant, the strength of the adsorption. Some articles [31,32] used the following linear equation, which is rearranged from Equation (4):

$$\lg(V) = n\cdot\lg(P) + \lg(K) \tag{5}$$

Additionally, the log-log plot of V versus P should be a straight line with the slope of n and the y-intercept of lg(K).

3.3. Absolute Adsorption Amount

The adsorption amount through a measurement test can be defined as excess adsorption amount, and the absolute adsorption amount can be converted as the following equation [33]:

$$n_{abs} = \frac{n_{excess}}{1 - \rho_{gas}/\rho_{ads}} \tag{6}$$

where n_{abs} is the absolute amount of methane adsorption, cc/g; n_{excess} is the excess amount of methane adsorption, cc/g; ρ_{gas} is the density of free phase gas, g/cc; and ρ_{ads} is the density of adsorbed phase gas in g/cc, which is determined as the value of 0.527 g/cc [34] and used in this study.

4. Results and Discussion

4.1. Dynamic Characteristics of Methane Adsorption

Figures 2–4 show the plots of pressure response and adsorption versus time under different pressure step at 30, 40 or 50 °C, respectively. Obviously, it is shown that the pressure drops fast at first, then slowly in the middle and reaches and maintains stability at last. Meanwhile, the absolute amount of methane adsorption increases quickly initially, then slowly in the middle and reaches a constant at last. These two curves, which describe the adsorption dynamic of methane, have similar characteristics to the other studies [8,24], indicating that the methane molecules are mainly adsorbed in the initial stage.

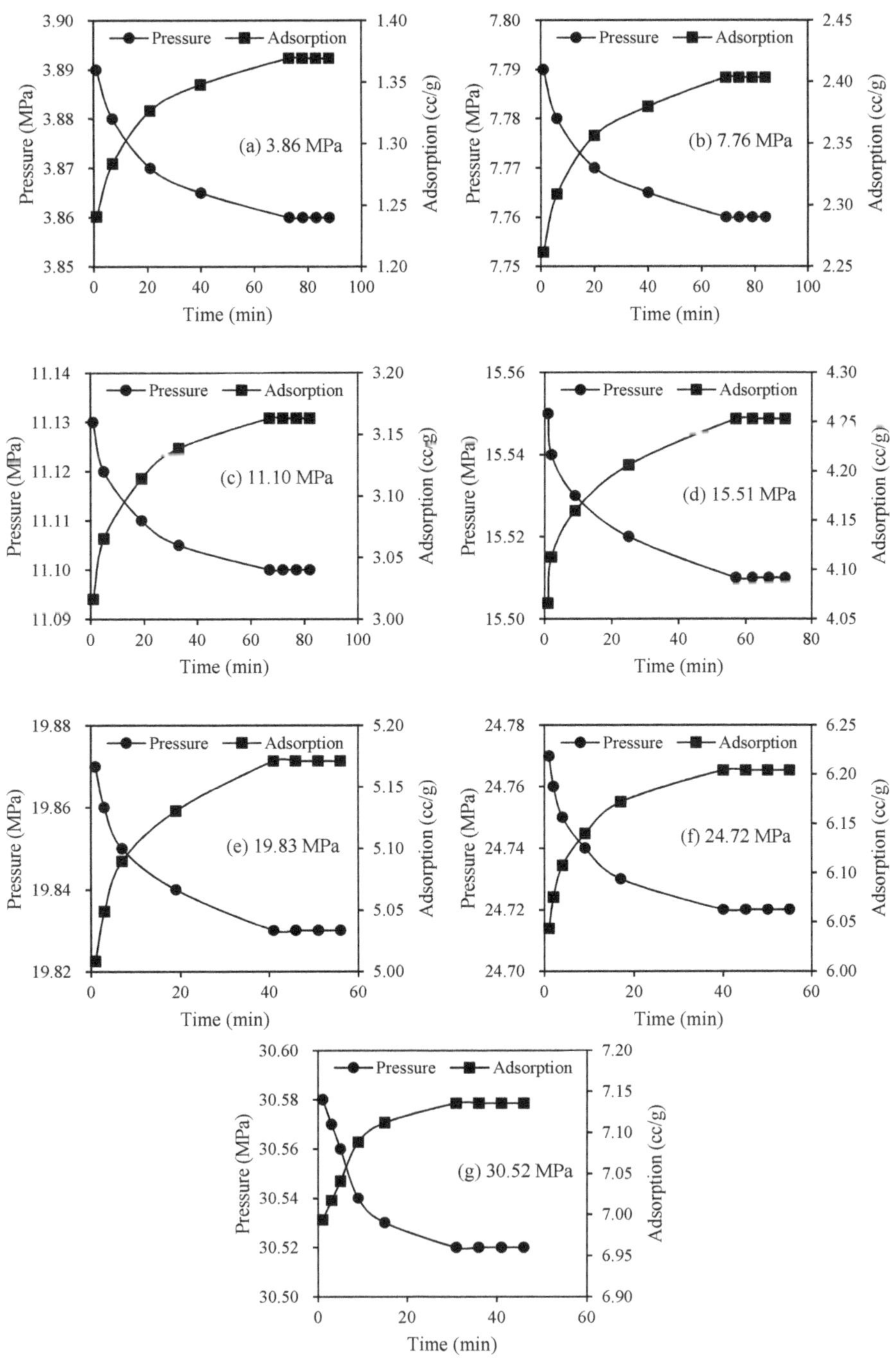

Figure 2. Methane adsorption dynamic on shale sample at seven pressure steps and 30 °C.

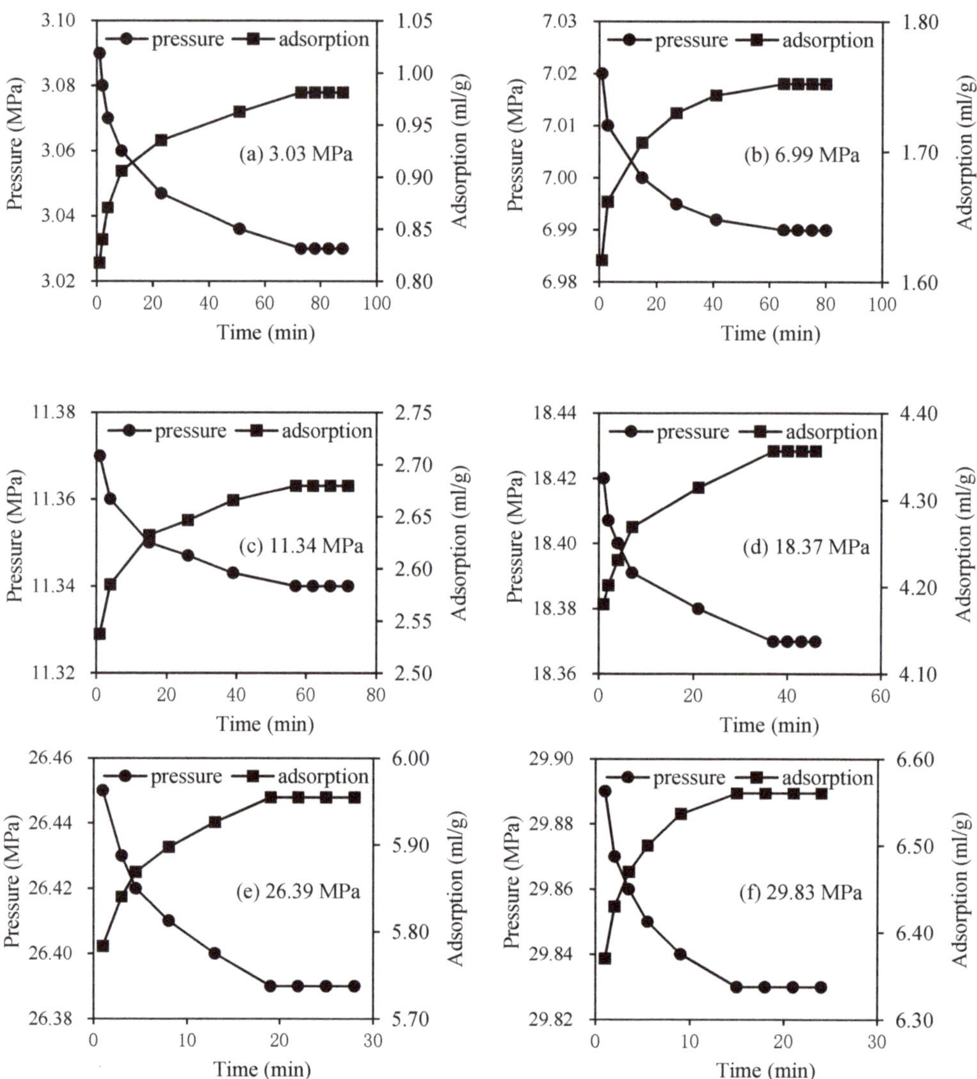

Figure 3. Methane adsorption dynamic on shale sample at six pressure steps and 40 °C.

To be more specific, at the beginning, most of the active surface sites are vacant and favorable for methane molecules adsorption on shale powders because the adsorption rate is positively correlated with the available vacancies [25]. Furthermore, at first, the high concentration driving force between the methane molecules spurs the mass free phase gas to the adsorbed phase gas. Additionally, in the middle, the repulsion of methane molecules gradually becomes the major force to determine the adsorption dynamic of methane molecules [35]. Therefore, the tendency of the adsorption dynamic curves demonstrates the synthetic influence of the high concentration driving force and repulsion of methane molecules on the adsorption dynamic of methane molecules.

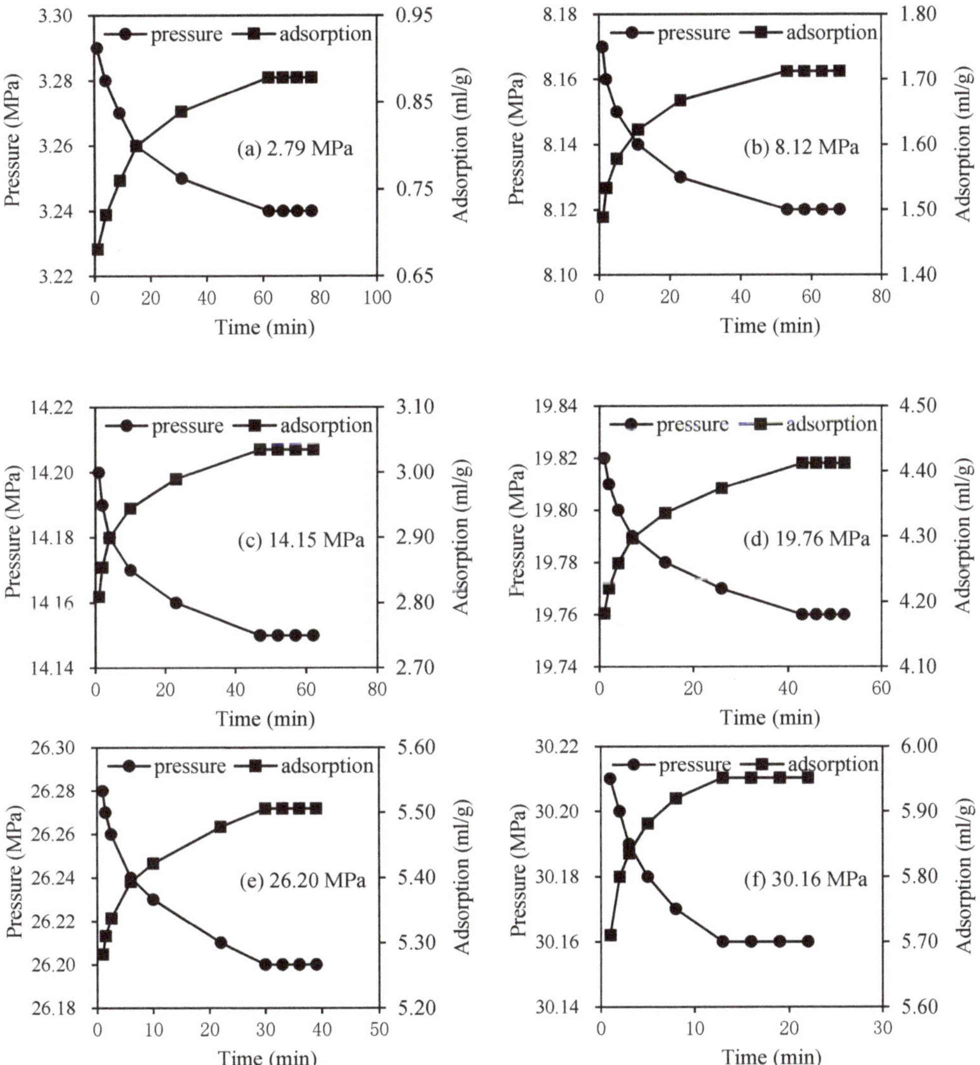

Figure 4. Methane adsorption dynamic on shale sample at six pressure steps and 50 °C.

4.2. Dynamic Model Fitting

Figure 5 shows the continuous change of $\ln\left(-\ln\left(\frac{q_e - q_t}{q_e}\right)\right)$ versus $\ln(t)$ under seven pressure steps using Equation (3) based on the Bangham model at 30 °C. It is clearly shown that $\ln\left(-\ln\left(\frac{q_e - q_t}{q_e}\right)\right)$ increases linearly with the increasing $\ln(t)$ because the correlation coefficients (R^2), respectively, are 0.9253, 0.9418, 0.9435, 0.9745, 0.9706, 0.9655 and 0.8585, as listed in Table 1, with the average of 0.9400. Therefore, q_t can be well fitted with t by using the Bangham model at each pressure step.

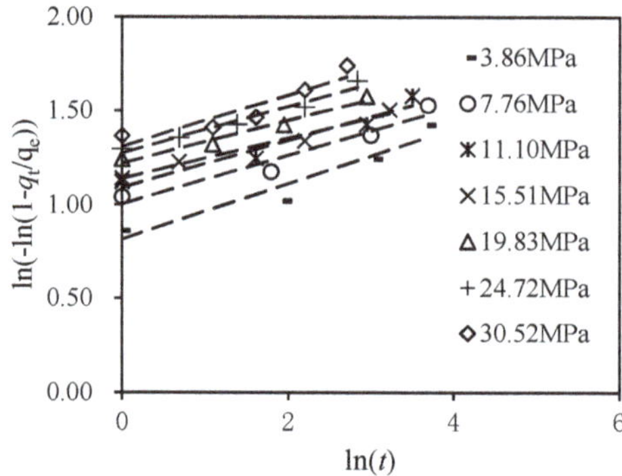

Figure 5. Fitting methane adsorption dynamic using Bangham model at 30 °C.

Table 1. The results of fitting methane adsorption dynamic using Bangham model at 30 °C.

NO	P (MPa)	Fitting Equation	R^2	Model	k	z
1	3.86	$y = 0.1475x + 0.8168$	0.9253	$q_t = 1.3694(1 - e^{-2.2632t^{\wedge}0.1457})$	2.2632	0.1457
2	7.76	$y = 0.1295x + 1.0038$	0.9418	$q_t = 2.4034(1 - e^{-2.7286t^{\wedge}0.1295})$	2.7286	0.1295
3	11.10	$y = 0.1259x + 1.0906$	0.9435	$q_t = 3.1632(1 - e^{-2.9761t^{\wedge}0.1259})$	2.9761	0.1259
4	15.51	$y = 0.1076x + 1.1385$	0.9745	$q_t = 4.2530(1 - e^{-3.1221t^{\wedge}0.1076})$	3.1221	0.1076
5	19.83	$y = 0.1145x + 1.2193$	0.9706	$q_t = 5.1712(1 - e^{-3.3848t^{\wedge}0.1145})$	3.3848	0.1145
6	24.72	$y = 0.1244x + 1.2742$	0.9655	$q_t = 6.2041(1 - e^{-3.5758t^{\wedge}0.1244})$	3.5758	0.1244
7	30.52	$y = 0.1379x + 1.3087$	0.8585	$q_t = 7.1357(1 - e^{-3.7014t^{\wedge}0.1379})$	3.7014	0.1379

The fitting equations, the Bangham model, the adsorption rate constant and the constant z are listed in Table 1. At different pressure steps, the adsorption rate constants are 2.2632, 2.7286, 2.9761, 3.1221, 3.3848, 3.5758 and 3.7014, respectively, indicating that the adsorption rate constant increases with the equilibrium pressure increasing. The constant z, respectively, is 0.1457, 0.1295, 0.1259, 0.1076, 0.1145, 0.1244 and 0.1379, revealing that z (a constant of the Bangham model) first decreases and then increases with the equilibrium pressure increasing.

Figure 6 shows the relationship between $\ln\left(-\ln\left(\frac{q_e - q_t}{q_e}\right)\right)$ and ln(t) under six pressure steps by using Equation (3) at 40 °C. It can be seen that $\ln\left(-\ln\left(\frac{q_e - q_t}{q_e}\right)\right)$ increases linearly with ln(t) increasing because the correlation coefficients at each pressure step, respectively, are 0.9938, 0.9430, 0.9411, 0.9600, 0.9512 and 0.9355, as listed in Table 2, with the average of 0.9541. Therefore, the Bangham model can be well fitted in the relationship between q_t and t for different pressure steps.

Table 2. The results of fitting methane adsorption dynamic using Bangham model at 40 °C.

NO	P (MPa)	Fitting Equation	R^2	Model	K	z
1	3.03	$y = 0.1993x + 0.5350$	0.9938	$q_t = 0.9809(1 - e^{-1.7074t^{\wedge}0.1993})$	1.7074	0.1993
2	6.99	$y = 0.1795x + 0.9028$	0.9430	$q_t = 1.7522(1 - e^{-2.4665t^{\wedge}0.1795})$	2.4665	0.1795
3	11.34	$y = 0.1469x + 1.0387$	0.9411	$q_t = 2.6797(1 - e^{-2.8255t^{\wedge}0.1469})$	2.8255	0.1469
4	18.37	$y = 0.1240x + 1.1319$	0.9600	$q_t = 4.3562(1 - e^{-3.1015t^{\wedge}0.1240})$	3.1015	0.1240
5	26.39	$y = 0.1544x + 1.2356$	0.9512	$q_t = 5.9546(1 - e^{-3.4404t^{\wedge}0.1544})$	3.4404	0.1544
6	29.83	$y = 0.2012x + 1.2358$	0.9355	$q_t = 6.5604(1 - e^{-3.4411t^{\wedge}0.2012})$	3.4411	0.2012

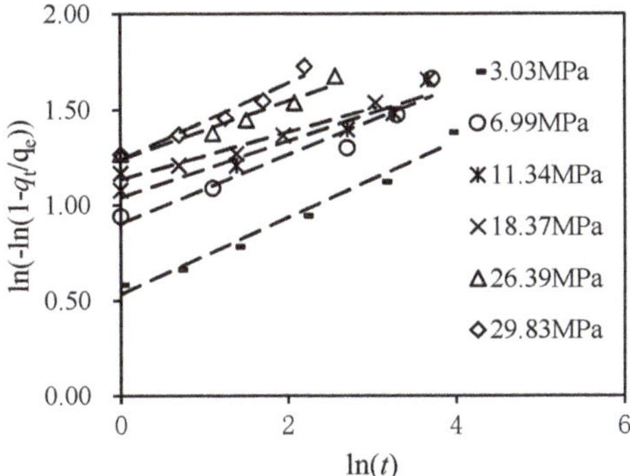

Figure 6. Fitting methane adsorption dynamic using Bangham model at 40 °C.

The fitting results, including the fitting equations, the Bangham model, the adsorption rate constant and the constant z are listed in Table 2. At six pressure steps, the adsorption rate constants are 1.7074, 2.4665, 2.8255, 3.1015, 3.4404 and 3.4411, respectively, indicating that k (adsorption rate constant) increases with the equilibrium pressure increasing. The constant z, respectively, is 0.1993, 0.1795, 0.1469, 0.1240, 0.1544 and 0.2012, revealing that z (a constant of the Bangham model) first decreases and then increases with the equilibrium pressure increasing.

Figure 7 shows the plots of $\ln\left(-\ln\left(\frac{q_e-q_t}{q_e}\right)\right)$ and ln(t) under six pressure steps using Equation (3) at 50 °C. It is clearly shown that $\ln\left(-\ln\left(\frac{q_e-q_t}{q_e}\right)\right)$ increases linearly with ln(t) increasing because the correlation coefficients at each pressure step, respectively, are 0.9170, 0.9743, 0.9864, 0.9610, 0.9411 and 0.9964, as listed in Table 3, with the average of 0.9627. Thus, q_t can be well fitted with t by using the Bangham model for six pressure steps.

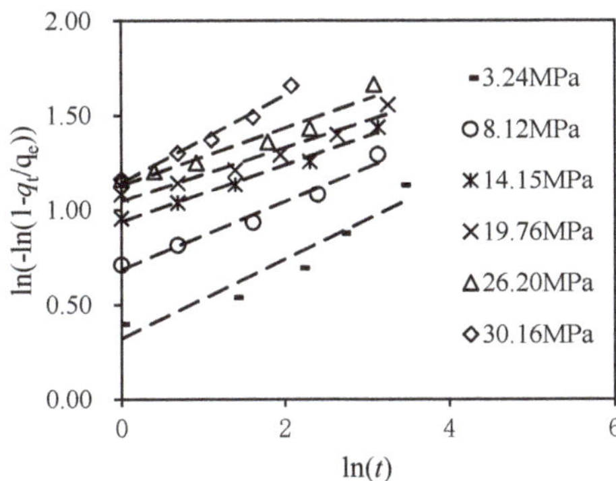

Figure 7. Fitting methane adsorption dynamic using Bangham model at 50 °C.

Table 3. The results of fitting methane adsorption dynamic using Bangham model at 50 °C.

NO	P(MPa)	Fitting Equation	R^2	Model	k	z
1	3.24	$y = 0.2086x + 0.3231$	0.9170	$q_t = 0.8787(1 - e^{-1.3814t^{\wedge}0.2086})$	1.3814	0.2086
2	8.12	$y = 0.1791x + 0.6864$	0.9743	$q_t = 1.7123(1 - e^{-1.9866t^{\wedge}0.1791})$	1.9866	0.1791
3	14.15	$y = 0.1508x + 0.9378$	0.9864	$q_t = 3.035(1 - e^{-2.5544t^{\wedge}0.1508})$	2.5544	0.1508
4	19.76	$y = 0.1415x + 1.0473$	0.9610	$q_t = 4.4132(1 - e^{-2.8499t^{\wedge}0.1415})$	2.8499	0.1415
5	26.20	$y = 0.1514x + 1.1314$	0.9411	$q_t = 5.206(1 - e^{-3.1000t^{\wedge}0.1314})$	3.1000	0.1514
6	30.16	$y = 0.1932x + 1.1660$	0.9964	$q_t = 5.9513(1 - e^{-3.2091t^{\wedge}0.1932})$	3.2091	0.1932

Table 3 lists the fitting results, including the fitting equations, the Bangham model, the adsorption rate constant and the constant z. K (adsorption rate constant) at six pressure steps, respectively, is 1.3814, 1.9866, 2.5544, 2.8499, 3.1000 and 3.2091, indicating that k increases with the equilibrium pressure increasing. The constant z (a constant of the Bangham model), respectively, is 0.2086, 0.1791, 0.1508, 0.1415, 0.1514 and 0.1932, revealing that z first decreases and then increases with the equilibrium pressure increasing.

4.3. *Effect of Temperature on Constant z*

To investigate the temperature influence on the adsorption dynamic of methane on shale powders, the constant z of the Bangham model at 30, 40 and 50 °C is plotted versus the equilibrium pressure, as shown in Figure 8. Obviously, it is shown that the curves of the constant z at different temperature conditions have a similar tendency to decrease first and then increase with the equilibrium pressure increasing. Furthermore, at the same pressure point, the higher the temperature, the smaller the constant z, indicating that temperature can obviously reduce the constant z. Therefore, the constant z first decreases and then increases with the equilibrium pressure increasing at each temperature, and it decreases with the temperature increasing at the same pressure.

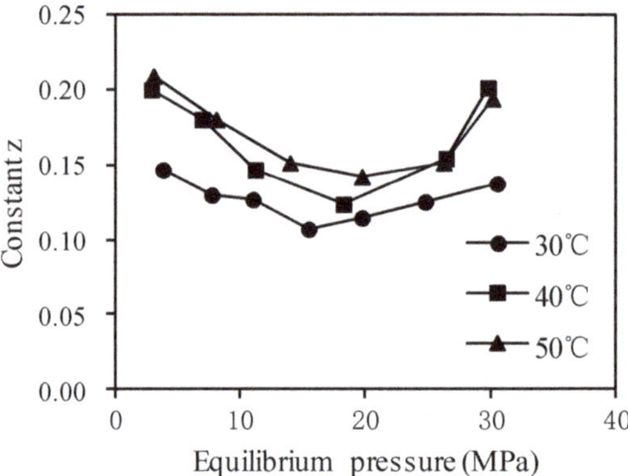

Figure 8. Constant Z versus equilibrium pressure at different temperature.

4.4. *Effect of Temperature on Adsorption Rate Constant*

The plot of the adsorption rate constant versus equilibrium pressure at 30, 40 and 50 °C is shown in Figure 9 to illustrate the temperature effect on the adsorption dynamic of methane molecules on shale powders. Obviously, it is shown that the adsorption rate constants all slowly drop with the equilibrium pressure increasing under different temperature conditions, revealing that it is much easier for methane molecules to adsorb on gas shale powders at a higher pressure. This is mainly because at a higher pressure condition, the high concentration driving force is the main controlling force that can promote the adsorption rate of methane molecules. Moreover, at the same pressure point, a smaller Bangham adsorption rate constant is attained due to a higher temperature, which indicates low temperatures are favorable for methane adsorption on shale powders. This is mainly because the methane adsorption dynamic on shale powders is exothermic.

Furthermore, to quantitatively analyze the relationship between the adsorption rate constant and the equilibrium pressure, the linear correlation relationships between k (adsorption rate constant) and $\ln(P)$ at 30, 40 and 50 °C are plotted in a semi-logarithmic coordinate system, as shown in Figure 10. The fitted results are, respectively, expressed as follows:

$$30\ ^\circ\mathrm{C} \qquad k = 0.6973\ln(P) + 1.2983 \qquad R^2 = 0.9932 \tag{7}$$

$$40\ ^\circ\mathrm{C} \qquad k = 0.7583\ln(P) + 0.9269 \qquad R^2 = 0.9924 \tag{8}$$

$$50\ ^{\circ}\mathrm{C} \qquad k = 0.8395\ln(P) + 0.3343 \qquad R^2 = 0.9937 \tag{9}$$

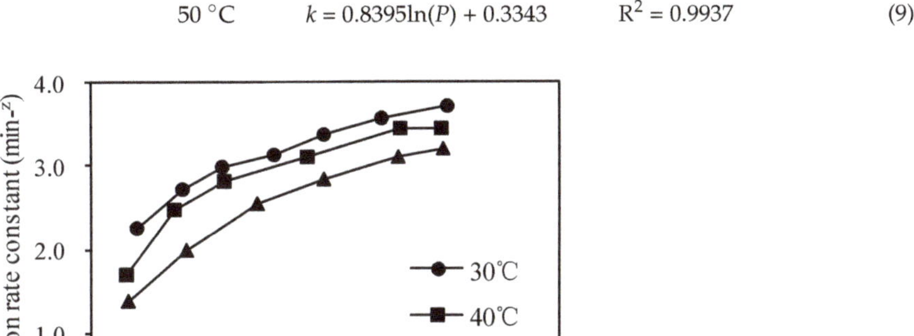

Figure 9. Adsorption rate constant versus equilibrium pressure at different temperature.

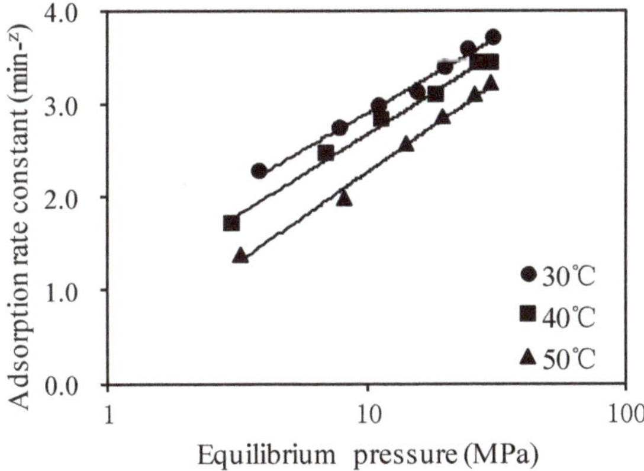

Figure 10. The relationship between adsorption rate constant and equilibrium pressure.

The correlation coefficients at 30, 40 and 50 °C, respectively, are 0.9932, 0.9924 and 0.9937, indicating that the adsorption rate constant is linearly positively correlated with the natural log of the equilibrium pressure.

4.5. Effect of Temperature on Adsorption Isotherm

To compare the difference in the adsorption amount at different temperatures, the adsorption amounts under different equilibrium pressures at 30, 40 and 50 °C are plotted versus the equilibrium pressure in Figure 11. It is clearly shown that the adsorption amount under the different equilibrium pressure at 30 °C is the biggest, followed by that of 40 °C and 50 °C, which indicates low temperatures are favorable for methane adsorption on shale powders. Figure 12 shows the plot of adsorption amount versus equilibrium pressure in a logarithmic coordinate system at each stable temperature. Obviously, it is shown that the adsorption amount is linearly positively correlated with the equilibrium pressure. The fitted results, including the fitted equation, the correlation coefficient (R^2), the Freundlich model, the Freundlich constant K and the constant n, are listed in Table 4. The correlation coefficients at 30, 40 or 50 °C, respectively, are 0.9945, 0.9987 and 0.9925, indicating that the relationship between the adsorption amount and the equilibrium pressure can be well fitted by the Freundlich model. K (Freundlich constant) at 30, 40 and 50 °C, respectively, is

5.1487, 2.1062 and 1.7857, indicating that K decreases with the temperature increasing. The constant n, respectively, is 0.2182, 0.2120 and 0.1967, revealing that the constant n decreases with the temperature increasing. Therefore, low temperatures are favorable for methane adsorption on shale powders, and high temperatures can obviously reduce constant K and n of the Freundlich model.

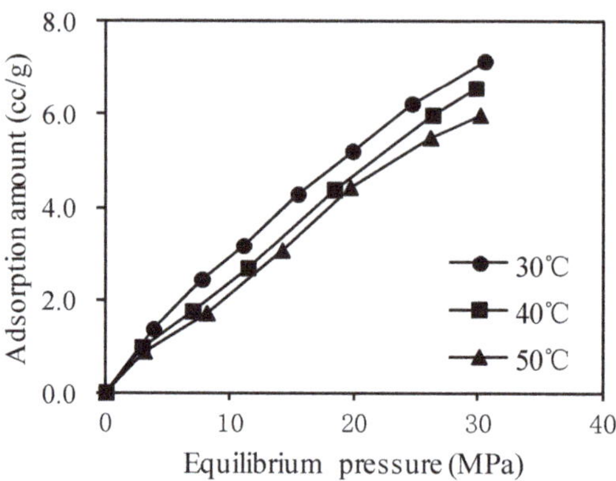

Figure 11. Adsorption amount versus equilibrium pressure at different temperature.

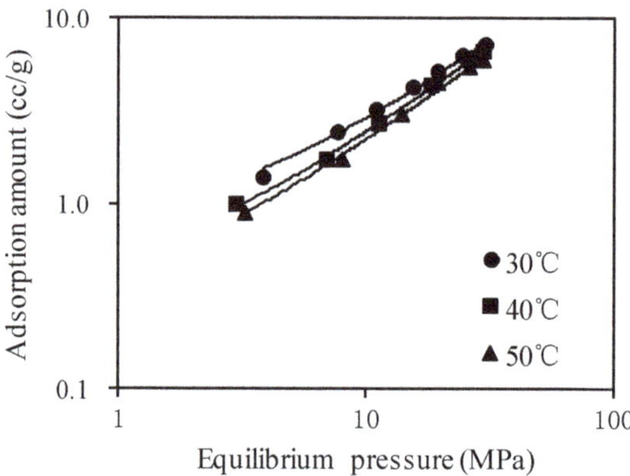

Figure 12. The relationship between adsorption amount and equilibrium pressure.

Table 4. The results of fitting methane adsorption isotherm at different temperatures.

NO	T (°C)	Fitting Equation	R^2	Freundlich Model	K	n
1	30	$y = 0.2182x + 0.7117$	0.9945	$V = 5.1484P^{0.2182}$	5.1487	0.2182
2	40	$y = 0.2120x + 0.3235$	0.9987	$V = 2.1062P^{0.2120}$	2.1062	0.2120
3	50	$y = 0.1967x + 0.2518$	0.9925	$V = 1.7857P^{0.1967}$	1.7857	0.1967

4.6. Effect of Temperature on Isostatic Enthalpy

The isostatic enthalpy of methane adsorption is derived from the Van't Hoff equation, and it is expressed as follows [16]:

$$\left(\frac{\partial \ln P}{\partial T}\right)_n = \frac{\Delta H}{RT^2} \quad (10)$$

where P is the pressure in kPa, T is the temperature in K, n is the absolute adsorption amount, R is the ideal gas constant in kJ/mol, and $\triangle H$ is the enthalpy of adsorption in kJ/mol.

Equation (10) can be integrated and rearranged, and the linear form of this model can be expressed as

$$\ln P = a\frac{1}{T} + \mathrm{b} \tag{11}$$

where $a = -\frac{\Delta H}{\mathrm{R}}$, $\mathrm{b} = \frac{\Delta H}{\mathrm{R}}\frac{1}{\mathrm{T}_0} + \ln \mathrm{P}_0$. The plot of ln$P$ versus $1/T$ should be fitted as a linear equation, and then, ΔH can be calculated according to the slope of a.

Figure 13 shows the plot of ln$P$ versus n (adsorption amount) at 30, 40 or 50 °C. The fitted results, including the temperature, the fitting equation, the correlation coefficient (R^2) and the parameters of the fitted equation, are listed in Table 5. It is clearly shown that there exists a well-linear relationship between ln$P$ and n because the correlation coefficients, respectively, are 0.9471, 0.9142 and 0.9205, with the average of 0.9273. The slopes of the fitted equation increase with the temperature increasing. Moreover, by using the fitted equation listed in Table 5, the values of ln$P$ at different temperatures are calculated and shown in Table 6.

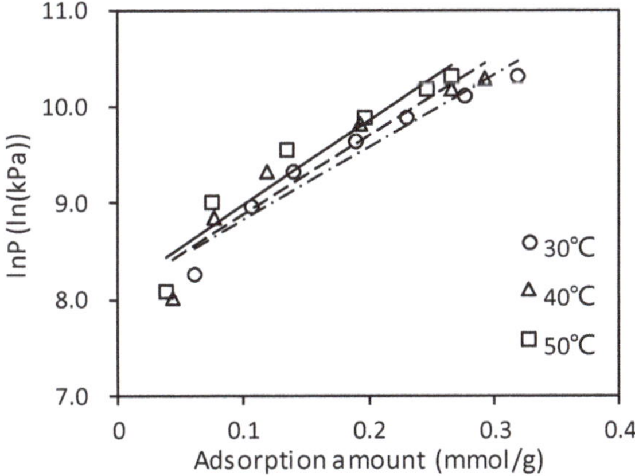

Figure 13. The relationship between ln$P$ and n.

Table 5. Fitting results of ln$P$ and n at different temperatures.

T (°C)	Fitting Equation	R^2	Slop	y-Intercept
30	$y = 7.5569x + 8.0708$	0.9471	7.5569	8.0708
40	$y = 8.2189x + 8.0551$	0.9142	8.2189	8.0551
50	$y = 8.7860x + 8.0984$	0.9205	8.7860	8.0984

Table 6. Calculation results of ln$P$ at different temperatures.

n (mmol/g)	ln$P$		
	30 °C	40 °C	50 °C
0.05	8.45	8.47	8.54
0.10	8.83	8.88	8.98
0.15	9.20	9.29	9.42
0.20	9.58	9.70	9.86
0.25	9.96	10.11	10.29
0.30	10.34	10.52	10.73
0.35	10.72	10.93	11.17

The relationship of ln$P$ and $1/T$ under different given adsorption amounts is shown in Figure 14, and the fitted results, including the adsorption amount, the fitted equation, the correlation coefficient (R^2) and the parameters of the fitted equation, are listed in Table 7. It can be seen that ln$P$ is linearly positively correlated with $1/T$ because R^2 is distributed between 0.8781 and 0.9974, with the average of 0.9705. Furthermore, isostatic enthalpy can be obtained, and the plot of isostatic enthalpy versus adsorption amount is shown in Figure 15. Obviously, it is shown that there exists a good linear relationship between isostatic enthalpy and the adsorption amount, indicating that isostatic enthalpy is linearly positively correlated with the adsorption amount.

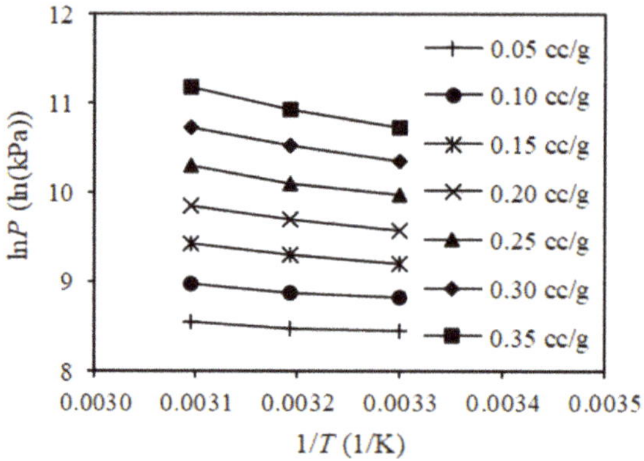

Figure 14. The relationship between ln$P$ and $1/T$ under different given adsorption amount.

Table 7. The results of fitting ln$P$ and $1/T$ under different given adsorption amount.

n (mmol/g)	Fitting Equation	R^2	a	b
0.05	$y = -433.23x + 9.87$	0.8781	−433.23	9.87
0.10	$y = -734.39x + 11.24$	0.9581	−734.39	11.24
0.15	$y = -1035.55x + 12.61$	0.9806	−1035.55	12.61
0.20	$y = -1336.71x + 13.98$	0.9895	−1336.71	13.98
0.25	$y = -1637.87x + 15.36$	0.9937	−1637.87	15.36
0.30	$y = -1939.03x + 16.73$	0.9961	−1939.03	16.73
0.35	$y = -2240.19x + 18.10$	0.9974	−2240.19	18.10

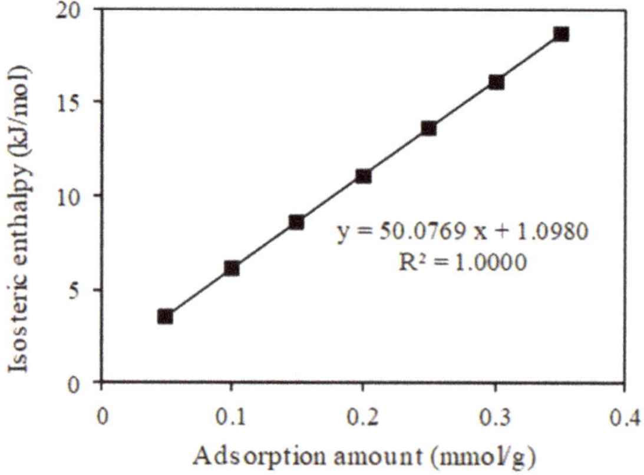

Figure 15. The relationship between isostatic enthalpy and adsorption amount.

5. Conclusions

The curves of the pressure response at different pressure steps and temperatures all have the same tendency to drop fast at first, then slowly in the middle and remain stable at last, and the adsorption amount increases quickly at first, then slowly in the middle and remains constant at last, revealing that the methane molecules are mainly adsorbed in the initial stage.

The adsorption amount (q_t) can be well fitted with time (t) by using the Bangham model at each pressure step. The constant z first decreases and then increases with the equilibrium pressure increasing at each temperature, and it decreases with the temperature increasing at the same pressure. The adsorption rate constant is linearly positively correlated with the natural log of the equilibrium pressure, and it decreases with the temperature increasing at the same pressure.

The Freundlich model can well fit the relationship between the adsorption amount and the equilibrium pressure. The constant K and n of the Freundlich model all decrease with the temperature increasing, indicating that the low temperature is favorable for methane adsorption on shale powders, and the high temperature can obviously reduce the constant K and n of the Freundlich model.

There exists a well-linear relationship between $\ln P$ and n at different temperatures. $\ln P$ is linearly positively correlated with $1/T$ under different given adsorption amounts. Isostatic enthalpy is linearly positively correlated with the adsorption amount.

Author Contributions: Conceptualization, H.Z. and Z.L.; methodology, Y.Z.; formal analysis, P.X.; data curation, A.H.; writing—original draft preparation, Y.Z.; writing—review and editing, Y.Z. and Z.L. All authors have read and agreed to the published version of the manuscript.

Funding: This research is supported by National Key Research and Development Program of China, grant number Sinopec Huabei Sub–Company Subcontract No. 2017ZX05049, and the Scientific and Technological Research Program of Chongqing Municipal Education Commission, grant number KJQN202101502.

Institutional Review Board Statement: Not applicable.

Informed Consent Statement: Not applicable.

Data Availability Statement: The data that support the findings of this study are available from the corresponding author upon reasonable request.

Conflicts of Interest: The authors declare no conflict of interest.

References

1. Ross, D.J.K.; Marc Bustin, R. Impact of mass balance calculations on adsorption capacities in microporous shale gas reservoirs. *Fuel* **2007**, *86*, 2696–2706. [CrossRef]
2. Curtis, J.B. Fractured shale-gas systems. *Am. Assoc. Pet. Geol. Bull.* **2002**, *86*, 1921–1938.
3. Wang, J.; Dong, M.; Yang, Z.; Gong, H.; Li, Y. Investigation of methane desorption and its effect on the gas production process from shale: Experimental and mathematical study. *Energy Fuels* **2017**, *31*, 205–216. [CrossRef]
4. Pan, Z.; Connell, L.D. Reservoir simulation of free and adsorbed gas production from shale. *J. Nat. Gas Sci. Eng.* **2015**, *22*, 359–370. [CrossRef]
5. Wang, J.; Wang, B.; Li, Y.; Yang, Z.; Gong, H.; Dong, M. Measurement of dynamic adsorption-diffusion process of methane in shale. *Fuel* **2016**, *172*, 37–48. [CrossRef]
6. Wu, F.C.; Tseng, R.L.; Juang, R.S. Initial behavior of intraparticle diffusion model used in the description of adsorption kinetics. *Chem. Eng. J.* **2009**, *153*, 1–8. [CrossRef]
7. Guo, C.; Li, R.; Sun, J.; Wang, X.; Liu, H. A review of gas transport and adsorption mechanisms in two-component methane-carbon dioxide system. *Int. J. Energy Res.* **2020**, *44*, 2499–2516. [CrossRef]
8. Dasani, D.; Wang, Y.; Tsotsis, T.T.; Jessen, K. Laboratory-Scale Investigation of Sorption Kinetics of Methane/Ethane Mixtures in Shale. *Ind. Eng. Chem. Res.* **2017**, *56*, 9953–9963. [CrossRef]
9. Chalmers, G.R.L.; Bustin, R.M. Lower Cretaceous gas shales in northeastern British Columbia, Part I: Geological controls on methane sorption capacity. *Bull. Can. Pet. Geol.* **2008**, *56*, 22–61. [CrossRef]

10. Manger, K.C.; Oliver, S.J.P.; Curtis, J.B.; Scheper, R.J. Geologic influences on the location and production of Antrim Shale gas, Michigan Basin. In Proceedings of the Rocky Mountain Regional/Low Permeability Reservoirs Symposium and Exhibition Rocky Mountain Regional/Low Permeability Reservoirs Symposium and Exhibition, Denver, CO, USA, 15–17 April 1991.
11. Heller, R.; Zoback, M. Adsorption of methane and carbon dioxide on gas shale and pure mineral samples. *J. Unconv. OIL GAS Resour.* **2014**, *8*, 14–24. [CrossRef]
12. Chen, J.; Wang, F.C.; Liu, H.; Wu, H.A. Molecular mechanism of adsorption/desorption hysteresis: Dynamics of shale gas in nanopores. *Sci. China Physics, Mech. Astron.* **2017**, *60*, 014611. [CrossRef]
13. Ross, D.J.K.; Marc Bustin, R. The importance of shale composition and pore structure upon gas storage potential of shale gas reservoirs. *Mar. Pet. Geol.* **2009**, *26*, 916–927. [CrossRef]
14. Zhang, T.; Ellis, G.S.; Ruppel, S.C.; Milliken, K.; Yang, R. Effect of organic-matter type and thermal maturity on methane adsorption in shale-gas systems. *Org. Geochem.* **2012**, *47*, 120–131. [CrossRef]
15. Fan, K.; Li, Y.; Elsworth, D.; Dong, M.; Yin, C.; Li, Y.; Chen, Z. Three stages of methane adsorption capacity affected by moisture content. *Fuel* **2018**, *231*, 352–360. [CrossRef]
16. Rexer, T.F.T.; Benham, M.J.; Aplin, A.C.; Thomas, K.M. Methane adsorption on shale under simulated geological temperature and pressure conditions. *Energy Fuels* **2013**, *27*, 3099–3109. [CrossRef]
17. Hu, A.; Zhang, Y.; Xiong, P.; Yang, Y.; Liu, Z. Kinetic characteristics and modeling comparison of methane adsorption on gas shale. *Energy Sources Part A Recover. Util. Environ. Eff.* **2020**, *00*, 1–15. [CrossRef]
18. Liu, Z.; Bai, B.; Wang, Y.; Ding, Z.; Li, J.; Qu, H.; Jia, Z. Experimental study of friction reducer effect on dynamic and isotherm of methane desorption on Longmaxi shale. *Fuel* **2021**, *288*, 119733. [CrossRef]
19. Meng, M.; Zhong, R.; Wei, Z. Prediction of methane adsorption in shale: Classical models and machine learning based models. *Fuel* **2020**, *278*, 118358. [CrossRef]
20. Plazinski, W.; Dziuba, J.; Rudzinski, W. Modeling of sorption kinetics: The pseudo-second order equation and the sorbate intraparticle diffusivity. *Adsorption* **2013**, *19*, 1055–1064. [CrossRef]
21. Guo, C.; Li, R.; Wang, X.; Liu, H. Study on two component gas transport in nanopores for enhanced shale gas recovery by using carbon dioxide injection. *Energies* **2020**, *13*, 1101. [CrossRef]
22. Wu, K.; Li, X.; Wang, C.; Yu, W.; Chen, Z. Model for surface diffusion of adsorbed gas in nanopores of shale gas reservoirs. *Ind. Eng. Chem. Res.* **2015**, *54*, 3225–3236. [CrossRef]
23. Yuan, W.; Pan, Z.; Li, X.; Yang, Y.; Zhao, C.; Connell, L.D.; Li, S.; He, J. Experimental study and modelling of methane adsorption and diffusion in shale. *Fuel* **2014**, *117*, 509–519. [CrossRef]
24. Chen, L.; Zuo, L.; Jiang, Z.; Jiang, S.; Liu, K.; Tan, J.; Zhang, L. Mechanisms of shale gas adsorption: Evidence from thermodynamics and kinetics study of methane adsorption on shale. *Chem. Eng. J.* **2019**, *361*, 559–570. [CrossRef]
25. Rani, S.; Prusty, B.K.; Pal, S.K. Adsorption kinetics and diffusion modeling of CH4 and CO2 in Indian shales. *Fuel* **2018**, *216*, 61–70. [CrossRef]
26. Yang, Z.; Wang, W.; Dong, M.; Wang, J.; Li, Y.; Gong, H.; Sang, Q. A model of dynamic adsorption-diffusion for modeling gas transport and storage in shale. *Fuel* **2016**, *173*, 115–128. [CrossRef]
27. Reveles, R.; Bangham, M.; Allen, J.; Schwartz, J.; Lambert, R.; Mangan, B. Characteristics of methane-oxygen combustion in explosions. In Proceedings of the 2018 Joint Propulsion Conference, Cincinnati, OH, USA, 9–11 July 2018.
28. Wang, Y.; Zhu, Y.; Liu, S.; Zhang, R. Methane adsorption measurements and modeling for organic-rich marine shale samples. *Fuel* **2016**, *172*, 301–309. [CrossRef]
29. Guo, J.; Zhai, Z.; Wang, L.; Wang, Z.; Wu, J.; Zhang, B.; Zhang, J. Dynamic and thermodynamic mechanisms of TFA adsorption by particulate matter. *Environ. Pollut.* **2017**, *225*, 175–183. [CrossRef]
30. Dada, A.O.; Olalekan, A.P.; Olatunya, A.M.; Dada, O.J. Langmuir, Freundlich, Temkin and Dubinin—Radushkevich Isotherms Studies of Equilibrium Sorption of Zn^{2+} Unto Phosphoric Acid Modified Rice Husk. *IOSR J. Appl. Chem.* **2012**, *3*, 38–45. [CrossRef]
31. Kanô, F.; Abe, I.; Kamaya, H.; Ueda, I. Fractal model for adsorption on activated carbon surfaces: Langmuir and Freundlich adsorption. *Surf. Sci.* **2000**, *467*, 131–138. [CrossRef]
32. Umpleby, R.J.; Baxter, S.C.; Bode, M.; Berch, J.K.; Shah, R.N.; Shimizu, K.D. Application of the Freundlich adsorption isotherm in the characterization of molecularly imprinted polymers. *Anal. Chim. Acta* **2001**, *435*, 35–42. [CrossRef]
33. Zhang, L.; Li, J.; Jia, D.; Zhao, Y.; Xie, C.; Tao, Z. Study on the adsorption phenomenon in shale with the combination of molecular dynamic simulation and fractal analysis. *Fractals* **2018**, *26*, 1–15. [CrossRef]
34. Zhou, S.; Xue, H.; Ning, Y.; Guo, W.; Zhang, Q. Experimental study of supercritical methane adsorption in Longmaxi shale: Insights into the density of adsorbed methane. *Fuel* **2018**, *211*, 140–148. [CrossRef]
35. Mall, I.D.; Srivastava, V.C.; Agarwal, N.K. Removal of Orange-G and Methyl Violet dyes by adsorption onto bagasse fly ash—Kinetic study and equilibrium isotherm analyses. *Dye. Pigment.* **2006**, *69*, 210–223. [CrossRef]

MDPI

Article

Study on Apparent Permeability Model for Gas Transport in Shale Inorganic Nanopores

Shuda Zhao [1], Hongji Liu [2], Enyuan Jiang [3], Nan Zhao [4], Chaohua Guo [2] and Baojun Bai [1,*]

1 GGPE, Missouri University of Science and Technology, Rolla, MO 65401, USA
2 Key Laboratory of Theory and Technology of Petroleum Exploration and Development, China University of Geosciences, Wuhan 430074, China
3 China National Oil and Gas Exploration and Development Co. Ltd., (CNOOC), Beijing 100034, China
4 SINOPEC Henan Oilfield Company, Nanyang 473132, China
* Correspondence: Baib@mst.edu

Abstract: Inorganic nanopores occurring in the shale matrix have strong hydrophilicity and irreducible water (IW) film can be formed on the inner surface of the pores making gas flow mechanisms in the pores more complex. In this paper, the existence of irreducible water (IW) in inorganic pores is considered, and, based on the Knudsen number (K_n) correction in shale pores, a shale gas apparent permeability model of inorganic nano-pores is established. The effect of the K_n correction on the apparent permeability, the ratio of flow with pore radius and the effect of IW on the apparent permeability are assessed. The main conclusions are as follows: (1) at low pressure (less than 10 MPa) and for medium pore size (pore radius range of 10 nm–60 nm), the effect of the K_n correction should be considered; (2) considering the effect of the K_n correction, bulk phase transport replaces surface diffusion more slowly; considering the existence of IW, bulk phase transport replaces surface diffusion more slowly; (3) with increase in pressure, the IW effect on gas apparent permeability decreases. Under low pressure, the IW, where pore size is small, promotes fluid flow, while the IW in the large pores hinders fluid flow. In conditions of ultra-high pressure, the IW promotes gas flow.

Keywords: shale gas; inorganic nanopore; apparent permeability; irreducible water; percolation mechanism; Knudsen number correction

Citation: Zhao, S.; Liu, H.; Jiang, E.; Zhao, N.; Guo, C.; Bai, B. Study on Apparent Permeability Model for Gas Transport in Shale Inorganic Nanopores. *Energies* **2022**, *15*, 6301. https://doi.org/10.3390/en15176301

Academic Editor: Hossein Hamidi

Received: 9 August 2022
Accepted: 27 August 2022
Published: 29 August 2022

1. Introduction

The world is facing serious environmental problems and energy shortages [1–3]. Countries around the world have gradually turned their attention to shale gas. A shale gas reservoir is an unconventional reservoir that integrates production, storage and accumulation [4–11]. The shale pore structure is very complex with a pore size in the nanometer range, and is very different from the pore structure of a conventional gas reservoir [12–17]. The complex pore structure determines the complex and variable storage status of shale gas and indirectly affects its production [18].

Beskok and Karniadakis [19] derived the Hagen–Poiseuille equation applicable at the nanometer scale and introduced a rare effect coefficient to the model. Javadpour [20] characterized the micro-migration characteristics of gas, establishing a gas mass flow equation similar to Darcy's formula, and first formally proposed the concept of apparent permeability. Civan [21] proposed a model for calculating apparent shale permeability using the Beskok–Karniadakis model, which considers the effects of curvature, inherent permeability, the gas slippage coefficient, porosity, and the rare effect coefficient on the gas flow. In this model, the parameters are relatively easy to obtain, so researchers often cite them. Xiong et al. [22] considered surface diffusion, studied the desorption of gas, and introduced a correction coefficient into the model showing that the gas mass flux was not represented by simple addition. For the first time, these authors proposed an apparent permeability model for shale gas transport which combined multiple mechanisms: surface diffusion,

Knudsen diffusion and viscous flow. Dababi et al. [23] extended the single circular tube nanopore Javadpour model to porous media and introduced the fractal dimension into Knudsen diffusion to address the influence of different pore surface roughness. Li et al. [24] studied the irreducible water (IW) in inorganic shale pores and established a gas transport model that considered the distribution of bound water in inorganic pores. Wu et al. [25] established a relatively complete apparent permeability model extending different properties of apparent permeability, which considers the real gas effect, the shale pore structure, the shake rock stress sensitivity and the matrix shrinkage effect. Wang et al. [26] introduced monolayer and multilayer adsorption models involving a variety of percolation mechanisms and developed a real gas flow model in organic shale pores.

However, there has been little research on the influence of water in shale pores on gas flow. In this paper, a shale gas apparent permeability model, considering the effect of irreducible water, is proposed. A correction of the Knudsen number in porous media is also considered. The influence of the Knudsen number correction is examined by comparing whether the Knudsen number correction is considered or not. The influence of irreducible water in pores on gas flow is analyzed by comparing whether the IW effect in pores is considered or not.

2. Establishment of Apparent Permeability Model

2.1. Assumptions

1. The composition of shale gas is considered to be methane only;
2. The occurrence of shale gas involves the coexistence of a free state and an adsorption state;
3. The change in pore radius caused by the presence of gas is considered;
4. Of the water molecules in the inorganic pores of shale gas, only irreducible water is considered—mobile water molecules are not considered;
5. The influence of stress on shale is not considered.

2.2. The Process of Permeability Model Establishment

Figure 1 represents the core idea of the paper and establishment of the mathematical model.

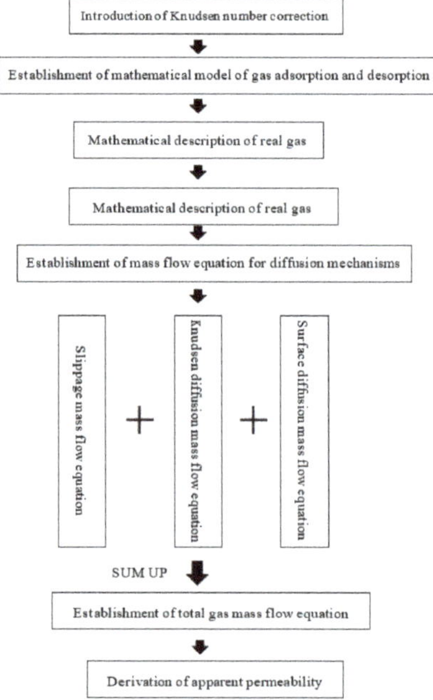

Figure 1. Flowchart of permeability model establishment.

2.3. Derivation of Gas Flow Equation

The Knudsen number in free space is generally calculated as follows [27]:

$$K_n = \frac{1}{2\sqrt{2}\pi N_g d^2 r} \tag{1}$$

where,

$$N_g = \frac{P}{k_b T} \tag{2}$$

N_g is the molecular density, m^3; d is the molecular diameter, m; k_b is the Boltzmann constant, 1.38×10^{-23} J/K; r is the pore radius, m; and P is the gas pressure, Pa.

In porous media, calculating the Knudsen number involves consideration of the intermolecular collisions and interactions with the media to produce additional collisions [28]. The Knudsen number in porous media can be expressed as:

$$K_{np} = \frac{\lambda_m}{2r_e} = \frac{1}{\left(2\sqrt{2}\pi N_g d^2 + \frac{\zeta S_r \rho_r}{2\phi}\right) r_e} \tag{3}$$

where, S_r is the shale specific surface area, m^2/g; ρ_r is the shale density, kg/m^3; ϕ is the porosity, dimensionless; and ζ is the equation coefficient, 1000.

In inorganic shale, the existence of water film should be considered. The three-phase adsorption and desorption model is used to calculate the gas adsorption volume [29]:

$$V = \beta \frac{V_H P}{P + P_H} + (1 - \beta) \frac{V_L P}{P + P_L} \tag{4}$$

where, V is the gas adsorption volume, m^3/kg; V_L is the Langmuir saturated adsorption volume, m^3/kg; P_L is the Langmuir pressure constant, Pa; V_H is the liquid-gas Langmuir adsorption volume, m^3/kg; P_H is the liquid-gas pressure constant, Pa; and β is the coverage of water molecules. The definition is as follows:

$$\beta = \frac{A_H}{A_{total}} \tag{5}$$

where, A_{total} is the total surface area of shale pores, m^2; and A_H is the surface area wetted by water molecules, m^2.

The total surface coverage of adsorbed gas in the three phase adsorption model is:

$$\theta = \frac{(1 - \beta)P}{P + P_L} + \frac{\beta P}{P + P_H} \tag{6}$$

In inorganic pores, the shale matrix pore radius is modified by the three-phase adsorption and desorption model as follows [30]:

$$r_e = r_m - d_H \beta - d\left[\frac{(1 - \beta)P}{P + P_L} + \frac{\beta P}{P + P_H}\right] \tag{7}$$

where, r_e is the real pore radius, m; d_H is the diameter of water molecules, m; d is the diameter of methane molecules, m; and r_m is the pore radius without fillers, m. Considering the real gas, the gas compression factor is as follows [31,32]:

$$Z = 0.702 P_r^2 e^{-2.5T_r} - 5.524 P_r e^{-2.5T_r} + 0.044 T_r^2 - 0.164 T_r + 1.15 \tag{8}$$

$$\begin{aligned} P_r &= P/P_c \\ T_r &= T/T_c \end{aligned} \tag{9}$$

where, P_r is the methane converted pressure, dimensionless; P_c is the methane critical pressure, Pa; T_r is the methane converted temperature, dimensionless; and T_c is the methane critical temperature, K. The equation for the equation for slippage mass flow with K_n is as follows [33]:

$$F_v = -\frac{\rho r_e^2}{8\mu}(1+\alpha \cdot K_n^*)\left(1+\frac{4K_n^*}{1-b\cdot K_n^*}\right)\nabla P \tag{10}$$

where, F_v is the slippage mass flow, kg/ (m^2·s); μ is the methane viscosity, Pa·s; ρ is the gas density, kg/m^3; and α is the fitting function about K_n. The expression is as follows [34]:

$$\alpha = \alpha_0 \tan^{-1}\left(4\cdot K_n^{*0.4}\right) \tag{11}$$

where, α_0 and β are fitting constants, dimensionless. Then the velocity form of the slippage flow is:

$$v_v = -\frac{r_e^2}{8\mu}(1+\alpha \cdot K_{np})\left(1+\frac{4K_{np}}{1-bK_{np}}\right)\nabla P \tag{12}$$

The Knudsen diffusion mass flow is as follows [35]:

$$F_k = -MD_k\nabla c = -\frac{MD_k}{ZRT}\nabla P \tag{13}$$

where, F_k is the Knudsen diffusion mass flow, kg/ (m^2·s); M is the molar mass of methane, kg/mol; and D_k is the Knudsen diffusion coefficient, m^2·s, which is given by the following equation:

$$D_k = \frac{2r_e}{3}\sqrt{\frac{8RT}{\pi M}} \tag{14}$$

Its flow velocity is:

$$v_k = -\frac{D_k}{P}\nabla P \tag{15}$$

The coefficient correction of the bulk phase transport mechanism is as follows [34]:

$$\begin{aligned} \varepsilon_v &= \frac{1}{1+K_n^*} \\ \varepsilon_k &= \frac{1}{1+1/K_n^*} \end{aligned} \tag{16}$$

Then the surface diffusion mass flow is as follows [36]:

$$F_s = -\frac{D_s C_s}{P}\nabla P \tag{17}$$

where, D_s is the gas surface diffusion coefficient, m^2/s; and C_s is the adsorption capacity of methane, kg/m^3, which is as follows:

$$C_s = \frac{4M\theta}{\pi d^3 N_A} \tag{18}$$

The surface diffusion coefficient is given as follows [37]:

$$D_s = D_{s0}\frac{(1-\theta)+\frac{\kappa}{2}\theta(2-\theta)+[H(1-\kappa)](1-\kappa)\frac{\kappa}{2}\theta^2}{\left(1-\theta+\frac{\kappa}{2}\theta\right)^2} \tag{19}$$

$$H(1-\kappa) = \begin{cases} 0, & \kappa \geq 1 \\ 1, & 0 \leq \kappa \leq 1 \end{cases} \tag{20}$$

where, D_{s_0} is the surface diffusion coefficient when $\theta = 0$, m^2/s. D_{s_0} is given as follows [38]:

$$D_s^0 = 8.29 \times 10^{-7} T^{0.5} \exp\left(-\frac{\Delta H_g^{0.8}}{RT}\right) \tag{21}$$

Considering the effect of irreducible water, $D_s 0$ is revised as follows [39]:

$$D_s^0 = 8.29 \times 10^{-7} T^{0.5} \exp\left(-\frac{\Delta H_w^{0.8}}{RT}\right) \tag{22}$$

where, ΔH_w is the equivalent heat of gas adsorption considering the existence of IW, J/mol, as follows:

$$\Delta H_w = \beta \Delta H_{gw} + (1-\beta)\Delta H(0) \tag{23}$$

where, $\Delta H(0)$ is the equivalent adsorption heat on the dry pore surface when $\theta = 0$, J/mol; and $\Delta H_g w$ is the equivalent adsorption heat on the gas-water interface when $\theta = 0$, J/mol.

The flow velocity of gas surface diffusion is as follows:

$$v_s = \frac{1}{\rho} F_s = -\frac{ZRTD_sC_s}{MP^2} \nabla PH(0) \tag{24}$$

The total flow velocity is:

$$\begin{aligned} v &= \left(\frac{r_e^2}{r_m^2}\right)(\varepsilon_v v_v + \varepsilon_k v_k) + \left(1 - \frac{r_e^2}{r_m^2}\right) v_s \\ &= -\left\{\left(\frac{r_e^2}{r_m^2}\right)\left[\frac{\varepsilon_v r_e^2}{8\mu}(1 + \alpha \cdot K_n^*)\left(1 + \frac{4K_n^*}{1 - b \cdot K_n^*}\right) + \frac{\varepsilon_k M D_k}{ZRT}\right] + \left(1 - \frac{r_e^2}{r_m^2}\right)\frac{ZRTD_sC_s}{MP^2}\right\}\nabla P \end{aligned} \tag{25}$$

2.4. Gas Apparent Permeability Model

The calculation equation of gas non-Darcy seepage, as proposed by Javadpour, is as follows:

$$v = -\frac{k_{app}}{\mu} \nabla P \tag{26}$$

By comparing Equation (25) with Equation (26), the equation of apparent permeability can be derived:

$$k_{app} = \left(\frac{r_e^2}{r_m^2}\right)\frac{\varepsilon_r r_e^2}{8}(1 + \alpha \cdot K_n^*)\left(1 + \frac{4K_n^*}{1 - b \cdot K_n^*}\right) + \left(\frac{r_e^2}{r_m^2}\right)\frac{\varepsilon_k \mu M D_k}{ZRT} + \left(1 - \frac{r_e^2}{r_m^2}\right)\frac{\mu ZRTD_sC_s}{MP^2} \tag{27}$$

2.5. Model Verification

To verify the proposed apparent permeability model, we compare the data in our paper with those in [39]. The apparent permeability varying with pressure is shown in Figure 2. As can be seen from Figure 2, the results in this paper closely match the reference sources, indicating the correctness of the model proposed.

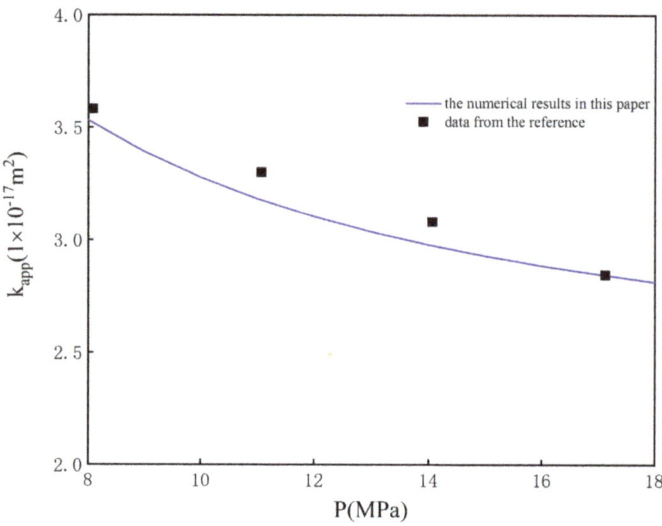

Figure 2. Comparison between the reference results and the numerical results in this paper.

3. Results and Disscussion

With the basic parameters of the solution, we can analyze the flow capacity of the $k_{a_{pp}}$ model, including the effect of K_{n_p} and the effect of irreducible water (IW). The basic parameters are shown in Table 1.

Table 1. List of basic parameters.

Parameter	Value and Unit	Meaning
d	0.38 nm	Methane molecular diameter
μ	0.0011 mPa·s	Methane gas viscosity
M	Methane gas viscosity	Methane molar mass
S_r	300 m^2/g	Specific surface area of shale
ρ_r	2.8×10^3 kg/m^3	Shale density
d_H	0.4 nm	Water molecular diameter
k_b	1.38×10^{-23} J/K	Boltzmann constant
R	8.314 J/mol/K	Ideal gas constant
P_L	2.1 MPa	Methane Langmuir pressure constant
α	1000	Unit conversion coefficient
ϕ	0.05	Shale porosity
b	−1	Slippage effect fitting constant
N_a	6.02×10^{23}	Avogadro constant
T	310 K	Temperature
Pc	4.539 MPa	Critical pressure of methane
Tc	190.5 K	Critical temperature of methane
$\Delta H_g w$	7953 J/mol	Gas-water interface when gas coverage is 0
$\Delta H(0)$	12,000 J/mol	Equivalent adsorption heat on the dry surface when gas coverage is 0
κ	0.5	Ratio of surface diffusion advance rate to blocking rate
P_H	22.28 MPa	Langmuir pressure constant of methane gas-water interface

3.1. Effect of Knudsen Number Correction

Taking r_m as 10 nm, the variation in apparent permeability under K_n or K_{n_p} with pressure is calculated, as shown in Figure 3.

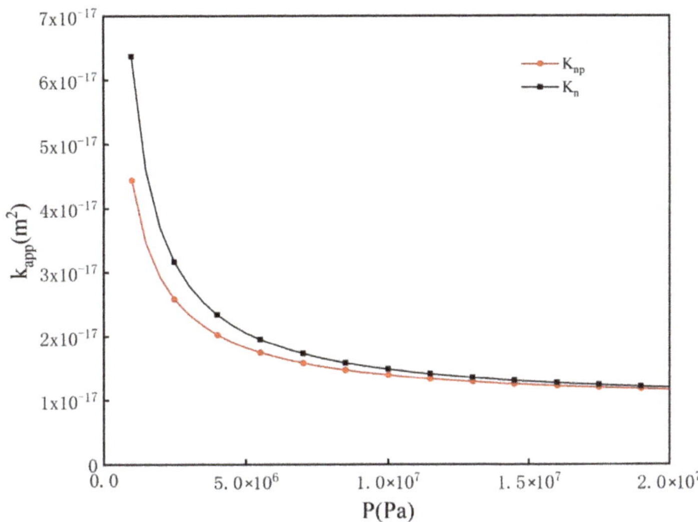

Figure 3. Apparent permeability vs. pressure with K_{n_p} or K_n (r_m = 10 nm).

It can be seen in Figure 3, that the apparent permeability with K_{n_p} is lower than that with K_n. The effect of K_{n_p} on the apparent permeability decreases when the pressure increases. After the pressure exceeds 10 MPa, this effect is almost negligible, so the K_n correction can be ignored. However, when the pressure is within 10 MPa, the K_n correction has a certain influence on the actual apparent permeability calculation, so it is necessary to consider the Knudsen correction in this situation. We calculate the change in apparent permeability with pore radius. At the same time, we take different pressures 1 MPa, 5 MPa, 10 MPa for comparative analysis. The result is shown in Figure 4.

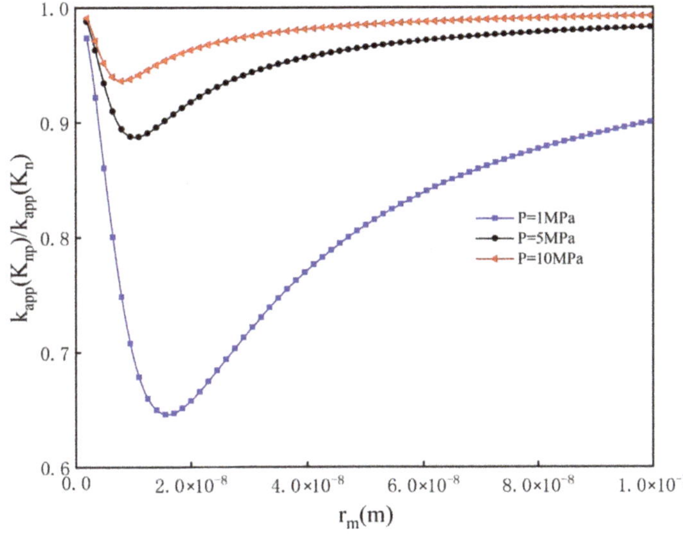

Figure 4. Apparent permeability ratio vs. pore radius with K_{n_p} or K_n under different pressures.

As seen in Figure 4, under different pressure, with increase in pore radius, the effect of K_{n_p} on apparent permeability first increases and then decreases, and the final value approaches 1. At the same time, the smaller the pressure is, the greater the peak value of the correction effect is. This is consistent with the earlier conclusion that the effect of the K_n correction is significant when the pressure is small. At the same time, it is noted that the pore radius range of the significant correction is that of medium pores. At medium pore size, about [10nm to 60nm], the effect of the K_n correction needs to be considered.

3.2. Flow Mechanism Analysis

The ratio of flow mechanism is defined as follows:

$$\gamma_1 = \frac{k_b}{k_{app}}; \gamma_2 = \frac{k_s}{k_{app}} \tag{28}$$

where, k_b is the apparent permeability of bulk transport, defined as follows:

$$k_b = k_v + k_k \tag{29}$$

Considering the existence of irreducible water (IW) or not, under 1 MPa and 10 MPa pressure, the extent of flow mechanism with pore radius is analyzed, as shown in Figures 5–8.

The following inferences can be drawn from the above figures:

(1) Under different conditions, the ratio of bulk phase transport to surface diffusion follows the same trend. With increase in the pore radius, the contribution of bulk phase gas transport also increases. With increase in the pore radius, the contribution of surface gas diffusion decreases, which is consistent with the nature of surface diffusion.

(2) Considering the effect of the K_n correction: If the ratio of bulk transport is equal to that of surface diffusion, the point of intersection is called the flow equivalent point. Comparing Figure 5 with Figure 6, and Figure 7 with Figure 8, it can be seen that, compared with considering K_n, the flow equivalent point considering K_{np} moves in the direction of increasing pore radius. This shows that, considering K_{n_p}, the rising trend in the bulk phase transport proportion and the decreasing trend in the surface diffusion proportion are smaller than those considering K_n. Moreover, bulk phase transport replaces surface diffusion more slowly than when considering K_n.

(3) When the pressure increases, the flow equivalent point moves in the direction of a small pore radius, which indicates that the rising trend in the bulk phase transport proportion, and the decreasing trend in the surface diffusion proportion, under high pressure, are larger than those under low pressure, and bulk phase transport replaces the surface diffusion faster.

(4) Considering the existence of irreducible water (IW), the flow equivalent point moves in the direction of a large pore radius, which indicates that the rising trend in the bulk phase transport proportion and the decreasing trend in the surface diffusion proportion are smaller than those without considering IW, and bulk phase transport replaces surface diffusion more slowly.

3.3. Effect of Irreducible Water

As can be seen from Figures 9 and 10, the gas apparent permeability decreases when the pressure increases. Under different pressure, the apparent permeability with a large β value is always larger than that with a low β value, and the difference under different conditions decreases when the pore radius increases.

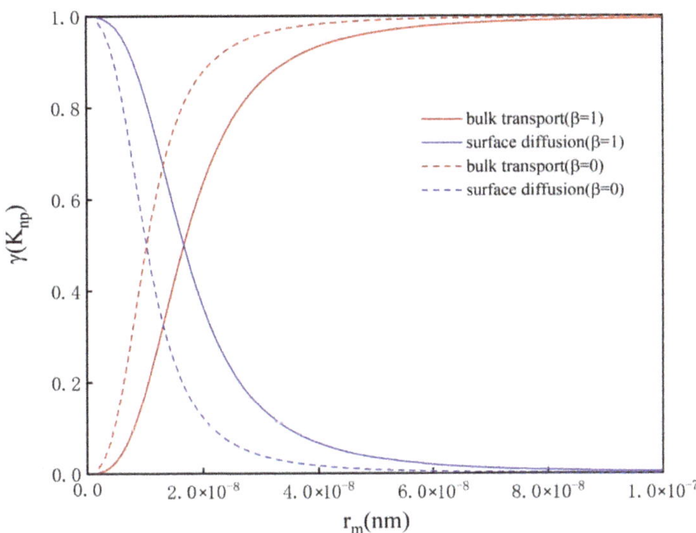

Figure 5. Under different degrees of water molecular coverage, considering K_{n_p}, the ratio of flow mechanism vs. pore radius (P = 1 MPa).

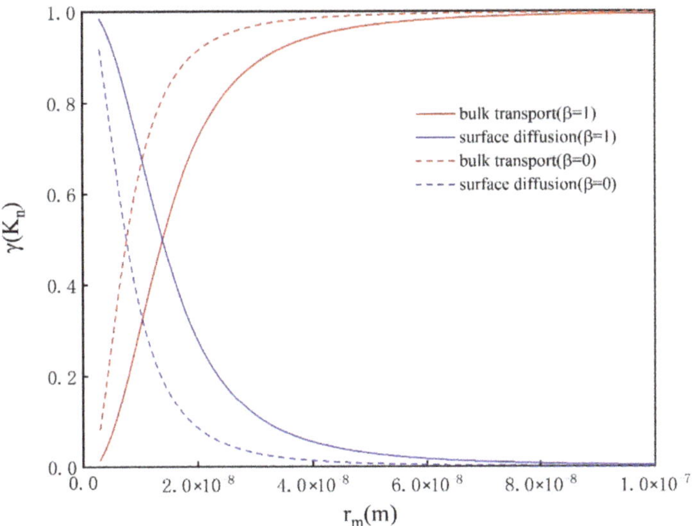

Figure 6. Under different degrees of water molecular coverage,considering K_n, the ratio of flow mechanism vs. pore radius (P = 1 MPa).

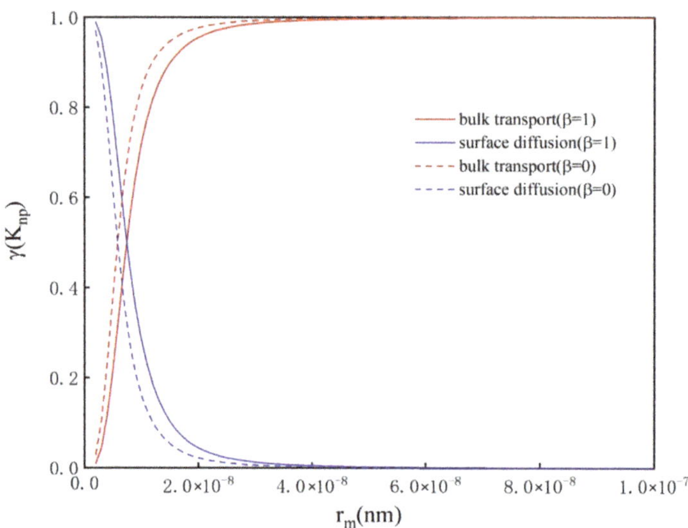

Figure 7. Under different degrees of water molecular coverage, considering K_{n_p}, the ratio of flow mechanism vs. pore radius (P = 10 MPa).

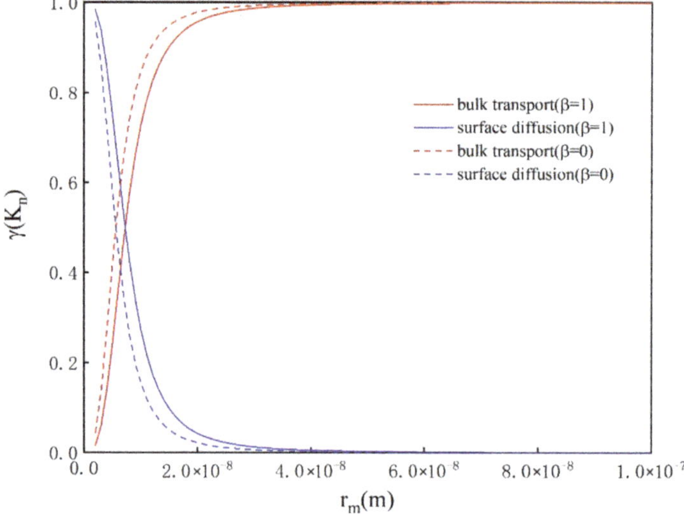

Figure 8. Under different degrees of water molecular coverage, considering K_n, the ratio of flow mechanism vs. pore radius (P = 10 MPa).

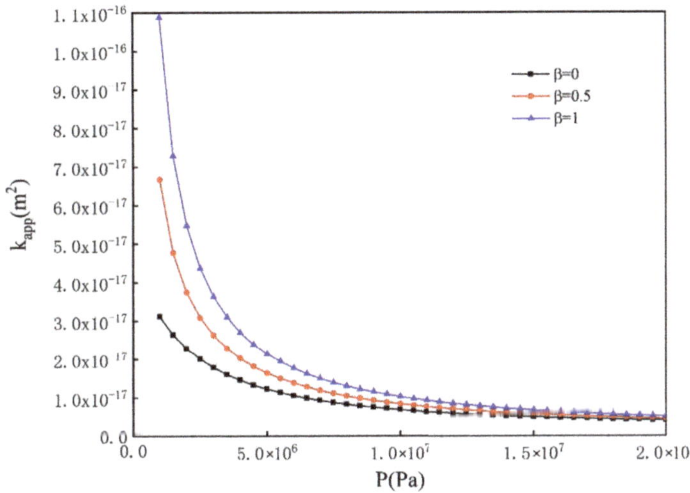

Figure 9. Apparent permeability vs. pressure under different water molecular coverage (r = 5 nm).

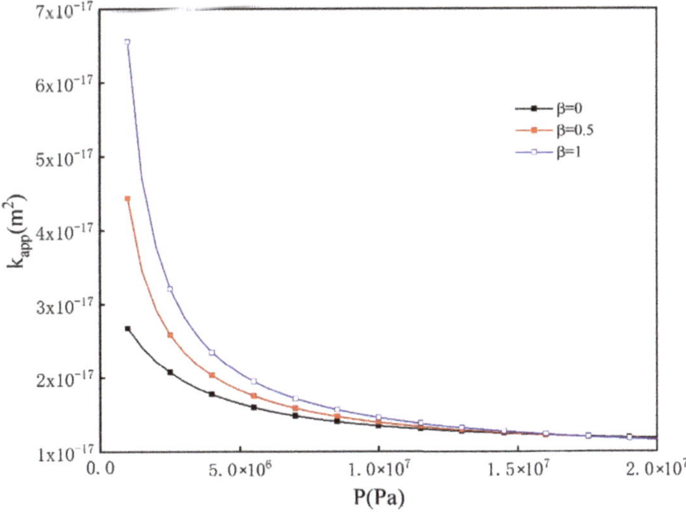

Figure 10. Apparent permeability vs. pressure under different water molecular coverage (r = 10 nm).

The characteristic parameter reflecting the effect of IW on gas transport in shale nanopores is defined as follows:

$$\eta = \frac{k_{app}(\beta = 0)}{k_{app}(\beta = 1)} \tag{30}$$

where, η is the ratio of the apparent permeability when the water molecule coverage is 0 to the apparent permeability when the water molecule coverage is 1. If $\eta > 1$, it indicates that the IW effect has a negative influence; that is, the existence of IW can hinder the gas flow

The variation in η with r_m under different pressure is shown in Figure 10.

It can be seen from Figure 11 that, with increase in the pore radius (r), the characteristic parameters gradually increase, and the value is greater than 1, then the characteristic parameters begin to decrease, but the value is still greater than 1. From the definition of the characteristic parameters, we know that the positive influence of IW decreases with increase in r; that is, in small pores, the effect of bound water is to strongly promote gas flow. In large pores, the positive influence of IW becomes weaker, and can even become negative. With further increase in r, the negative influence begins to decrease, and finally tends to have no effect on the IW.

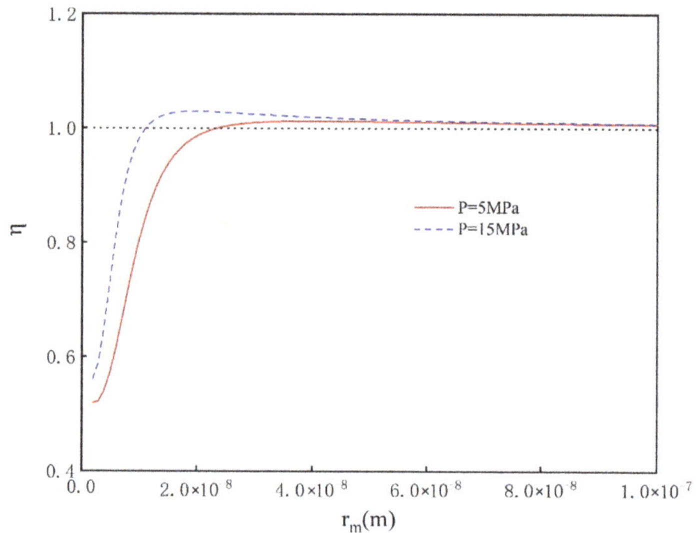

Figure 11. Characteristic parameters vs. pore radius under different pressures.

The effect of IW under low pressure is more obvious than that under high pressure, and the turning point is faster than that under high pressure. Under 5 MPa, the pore radius reaches about 10 nm, and the effect of IW begins to become negative; under 15 MPa, the pore radius reaches about 25 nm, and the effect of IW begins to turn into a negative effect. At the same time, according to the trend of the curve, although the effect of IW can be turned into a negative effect under ultra-high pressure, the negative effect is not obvious because the characteristic parameter of water content is too close to 1. Therefore, under ultra-high pressure, only the positive effect of IW is considered; that is, the effect of IW can promote gas flow under ultra-high pressure.

4. Conclusions

In inorganic pores, there are several flow mechanisms, such as Knudsen diffusion, the slippage effect and surface diffusion. At the same time, due to the existence of irreducible water (IW), the gas percolation process is more complex. The following conclusions are drawn from the study:

(1) Generally speaking, it is necessary to consider the Knudsen number correction in the medium pores (pore radius of about 10 nm to 60 nm) and under low pressure (within 10 MPa). However, under a high pressure and in small or large pores, the Knudsen number correction has a trivial effect on the gas apparent permeability.

(2) The ratio of bulk phase transport increases with increase in the pore radius, while the ratio of surface diffusion decreases with increase in the pore radius. Considering the effect of the Knudsen number correction, bulk phase transport replaces surface diffusion more slowly. Under high pressure, bulk phase transport replaces surface diffusion

faster. Considering the existence of IW, bulk phase transport replaces surface diffusion more slowly.

(3) The effect of IW on the gas apparent permeability decreases with increase in pressure. The existence of IW promotes fluid flow when the pore radius is smaller than the medium pore size (10 nm–60 nm). However, with increase in the pore radius, IW gradually exerts a negative influence on fluid flow. In the case of pressure lower than 0 MPa, the IW, where pore size is small, promotes fluid flow, while the IW, where pore size is large, hinders fluid flow. In the case of ultra-high pressure (larger than 15 MPa), the IW only promotes gas flow.

Author Contributions: Conceptualization, S.Z. and H.L.; methodology, S.Z.; software, C.G. and E.J.; validation, H.L., N.Z. and B.B.; formal analysis, S.Z. and H.L.; writing—original draft preparation, S.Z. and C.G.; writing—review and editing, S.Z. and C.G.; supervision, B.B. All authors have read and agreed to the published version of the manuscript.

Funding: This research was funded by National Natural Science Foundation of China: No. 51704265 and the Outstanding Talent Development Project of the China University of Geosciences (CUG20170614).

Institutional Review Board Statement: Not applicable.

Informed Consent Statement: Not applicable.

Data Availability Statement: Not applicable.

Acknowledgments: The authors acknowledge support from National Natural Science Foundation of China: No. 51704265 and the Outstanding Talent Development Project of the China University of Geosciences (CUG20170614).

Conflicts of Interest: The authors declare no conflict of interest.

References

1. Gamou, S.; Yokoyama, R.; Ito, K. Optimal unit sizing of cogeneration systems under the toleration for shortage of energy supply. *Int. J. Energy Res.* **2000**, *24*, 61–75. [CrossRef]
2. Li, J.; Li, M.; Li, Y. Strategies Analysis on Energy Shortage and Influence in China. In Proceedings of the 3rd Annual Meeting of Risk-Analysis-Council-of-China-Association-for-Disaster-Prevention, Guangzhou, China, 8–9 November 2008; pp. 729–734.
3. Xiao, Z.; Gao, J.; Wang, Z.; Yin, Z.; Xiang, L. Power shortage and firm productivity: Evidence from the World Bank Enterprise Survey. *Energy* **2022**, *247*, 123479. [CrossRef]
4. Gong, J.; Qiu, Z.; Zou, C.; Wang, H.; Shi, Z. An integrated assessment system for shale gas resources associated with graptolites and its application. *Appl. Energy* **2020**, *262*, 114524. [CrossRef]
5. Tonglou, G. Key geological issues and main controls on accumulation and enrichment of Chinese shale gas. *Pet. Explor. Dev.* **2016**, *43*, 349–359.
6. Guo, W.; Liu, H.L.; Xue, H.Q.; Li, X.B. Research and Applications on the Theory of Shale Gas Depth Enrichment Zone. In *Advanced Materials Research*; Trans Tech Publications: Wollerau, Switzerland, 2013; Volume 734, pp. 422–425.
7. Xusheng, G.; Dongfeng, H.; Yuping, L.; Zhihong, W.; Xiangfeng, W.; Zhujiang, L. Geological factors controlling shale gas enrichment and high production in Fuling shale gas field. *Pet. Explor. Dev.* **2017**, *44*, 513–523.
8. Kok, M.; Merey, S. Shale gas: Current perspectives and future prospects in Turkey and the world. *Energy Sources Part A Recovery Util. Environ. Eff.* **2014**, *36*, 2492–2501. [CrossRef]
9. Tang, X.; Jiang, Z.; Song, Y.; Luo, Q.; Li, Z.; Wang, G.; Wang, X. Advances on the mechanism of reservoir forming and gas accumulation of the Longmaxi Formation shale in Sichuan Basin, China. *Energy Fuels* **2021**, *35*, 3972–3988. [CrossRef]
10. Xie, J.; Qin, Q.; Fan, C. Influencing Factors of Gas Adsorption Capacity of Marine Organic Shale: A Case Study of Dingshan Area, Southeast Sichuan. *J. Coast. Res.* **2019**, *93*, 507–511. [CrossRef]
11. Xue, L.; Chen, B. Exploring on the measurement methods of shale gas content. In Proceedings of the 2016 3rd International Conference on Materials Engineering, Manufacturing Technology and Control, Taiyuan, China, 27–28 February 2016; Atlantis Press: Dordrecht, The Netherlands, 2016; pp. 575–578.
12. Haikuan, N.; Jinchuan, Z.; Shengling, J. Types and characteristics of the Lower Silurian shale gas reservoirs in and around the Sichuan Basin. *Acta Geol. Sin.-Engl. Ed.* **2015**, *89*, 1973–1985. [CrossRef]
13. Wang, Z. RETRACTED ARTICLE: Characterization of the microscopic pore structure of the lower paleozoic shale gas reservoir in the Southern Sichuan Basin and its influence on gas content. *Pet. Sci. Technol.* **2017**, *35*, 2165–2171. [CrossRef]
14. Yao, J.; Song, W.; Li, Y.; Sun, H.; Yang, Y.; Zhang, L. Study on the influence of organic pores on shale gas flow ability. *Sci. Sin. Phys. Mech. Astron.* **2017**, *47*, 094702. [CrossRef]

15. Zhang, J.; Li, X.; Zou, X.; Li, J.; Xie, Z.; Wang, F. Characterization of multi-type pore structure and fractal characteristics of the Dalong Formation marine shale in northern Sichuan Basin. *Energy Sources Part A Recovery Util. Environ. Eff.* **2020**, *42*, 2764–2777. [CrossRef]
16. Zhang, J.; Li, X.; Xiaoyan, Z.; Zhao, G.; Zhou, B.; Li, J.; Xie, Z.; Wang, F. Characterization of the full-sized pore structure of coal-bearing shales and its effect on shale gas content. *Energy Fuels* **2019**, *33*, 1969–1982. [CrossRef]
17. Zhang, P.; Hu, L.; Meegoda, J.N. Pore-scale simulation and sensitivity analysis of apparent gas permeability in shale matrix. *Materials* **2017**, *10*, 104. [CrossRef]
18. Chen, L.; Zuo, L.; Jiang, Z.; Jiang, S.; Liu, K.; Tan, J.; Zhang, L. Mechanisms of shale gas adsorption: Evidence from thermodynamics and kinetics study of methane adsorption on shale. *Chem. Eng. J.* **2019**, *361*, 559–570. [CrossRef]
19. Beskok, A.; Karniadakis, G.E. Report: A model for flows in channels, pipes, and ducts at micro and nano scales. *Microscale Thermophys. Eng.* **1999**, *3*, 43–77.
20. Javadpour, F. Nanopores and apparent permeability of gas flow in mudrocks (shales and siltstone). *J. Can. Pet. Technol.* **2009**, *48*, 16–21. [CrossRef]
21. Civan, F. Effective correlation of apparent gas permeability in tight porous media. *Transp. Porous Media* **2010**, *82*, 375–384. [CrossRef]
22. Xiong, X.; Devegowda, D.; Michel, G.; Sigal, R.F.; Civan, F. A fully-coupled free and adsorptive phase transport model for shale gas reservoirs including non-Darcy flow effects. In Proceedings of the SPE Annual Technical Conference and Exhibition, San Antonio, TX, USA, 8–10 October 2012. [CrossRef]
23. Darabi, H.; Ettehad, A.; Javadpour, F.; Sepehrnoori, K. Gas flow in ultra-tight shale strata. *J. Fluid Mech.* **2012**, *710*, 641–658. [CrossRef]
24. Li, J.; Chen, Z.; Wu, K.; Zhang, T.; Zhang, R.; Xu, J.; Li, R.; Qu, S.; Shi, J.; Li, X. Effect of water saturation on gas slippage in circular and angular pores. *AIChE J.* **2018**, *64*, 3529–3541. [CrossRef]
25. Wu, K.; Chen, Z.; Li, X.; Xu, J.; Li, J.; Wang, K.; Wang, H.; Wang, S.; Dong, X. Flow behavior of gas confined in nanoporous shale at high pressure: Real gas effect. *Fuel* **2017**, *205*, 173–183. [CrossRef]
26. Wang, J.; Kang, Q.; Wang, Y.; Pawar, R.; Rahman, S.S. Simulation of gas flow in micro-porous media with the regularized lattice Boltzmann method. *Fuel* **2017**, *205*, 232–246. [CrossRef]
27. Verma, B.; Demsis, A.; Agrawal, A.; Prabhu, S. Semiempirical correlation for the friction factor of gas flowing through smooth microtubes. *J. Vac. Sci. Technol. A Vacuum Surfaces Film.* **2009**, *27*, 584–590. [CrossRef]
28. Zeng, S.; Hunt, A.; Greif, R. Mean free path and apparent thermal conductivity of a gas in a porous medium. *J. Heat Transf.* **1995**, *117*. [CrossRef]
29. Li, J.; Li, X.; Wu, K.; Wang, X.; Shi, J.; Yang, L.; Zhang, H.; Sun, Z.; Wang, R.; Feng, D. Water sorption and distribution characteristics in clay and shale: Effect of surface force. *Energy Fuels* **2016**, *30*, 8863–8874. [CrossRef]
30. Guo, C.; Liu, H.; Xu, L.; Zhou, Q. An improved transport model of shale gas considering three-phase adsorption mechanism in nanopores. *J. Pet. Sci. Eng.* **2019**, *182*, 106291. [CrossRef]
31. Song, W.; Yao, J.; Li, Y.; Sun, H.; Zhang, L.; Yang, Y.; Zhao, J.; Sui, H. Apparent gas permeability in an organic-rich shale reservoir. *Fuel* **2016**, *181*, 973–984. [CrossRef]
32. Li, D.; Wang, Y.; Zeng, B.; Lyu, Q.; Zhou, X.; Zhang, N. Optimization of the shale gas reservoir fracture parameters based on the fully coupled gas flow and effective stress model. *Energy Sci. Eng.* **2021**, *9*, 676–693. [CrossRef]
33. Klinkenberg, L. The permeability of porous media to liquids and gases. *Am. Petrol. Inst. Drill. Prod. Pract.* **1941**, *2*, 200–213. [CrossRef]
34. Wu, K.; Li, X.; Guo, C.; Wang, C.; Chen, Z. A unified model for gas transfer in nanopores of shale-gas reservoirs: Coupling pore diffusion and surface diffusion. *SPE J.* **2016**, *21*, 1583–1611. [CrossRef]
35. Kast, W.; Hohenthanner, C.R. Mass transfer within the gas-phase of porous media. *Int. J. Heat Mass Transf.* **2000**, *43*, 807–823. [CrossRef]
36. Krishna, R.; Wesselingh, J. The Maxwell-Stefan approach to mass transfer. *Chem. Eng. Sci.* **1997**, *52*, 861–911. [CrossRef]
37. Chen, Y.; Yang, R.T. Surface and mesoporous diffusion with multilayer adsorption. *Carbon* **1998**, *36*, 1525–1537. [CrossRef]
38. Guo, L.; Peng, X.; Wu, Z. Dynamical characteristics of methane adsorption on monolith nanometer activated carbon. *J. Chem. Ind. Eng.* **2008**, *59*, 2726–2732.
39. Bai, J.; Kang, Y.; Chen, M.; Li, X.; You, L.; Chen, Z.; Fang, D. Impact of water film on methane surface diffusion in gas shale organic nanopores. *J. Pet. Sci. Eng.* **2021**, *196*, 108045. [CrossRef]

MDPI

Article

Cyclic Supercritical Multi-Thermal Fluid Stimulation Process: A Novel Improved-Oil-Recovery Technique for Offshore Heavy Oil Reservoir

Jie Tian *, Wende Yan *, Zhilin Qi, Shiwen Huang, Yingzhong Yuan and Mingda Dong

School of Petroleum and Natural Gas Engineering, Chongqing University of Science & Technology, Chongqing 401331, China

* Correspondence: tianjie@cqust.edu.cn (J.T.); 2012020@cqust.edu.cn (W.Y.)

Citation: Tian, J.; Yan, W.; Qi, Z.; Huang, S.; Yuan, Y.; Dong, M. Cyclic Supercritical Multi-Thermal Fluid Stimulation Process: A Novel Improved-Oil-Recovery Technique for Offshore Heavy Oil Reservoir. *Energies* **2022**, *15*, 9189. https://doi.org/10.3390/en15239189

Academic Editor: Daoyi Zhu

Received: 1 November 2022
Accepted: 28 November 2022
Published: 4 December 2022

Publisher's Note: MDPI stays neutral with regard to jurisdictional claims in published maps and institutional affiliations.

Abstract: Cyclic supercritical multi-thermal fluid stimulation (CSMTFS) is a novel technology that can efficiently recover heavy oil, while the heating effect, production and heat loss characteristics of CSMTFS have not been discussed. In this study, a physical simulation experiment of CSMTFS is conducted with a three-dimensional experimental system. The results of the study indicate that the whole process of CSMTFS can be divided into four stages, namely, the preheating stage, production increase stage, production stable stage and production decline stage, of which the production stable stage is the main oil production stage, and the production decline stage is the secondary oil production stage. In the first two stages of the CSMTFS process, there is no supercritical multi-thermal fluid chamber, and only a relatively small supercritical multi-thermal fluid chamber is formed in the last stage of the CSMTFS process. Out of the supercritical multi-thermal fluid chamber, supercritical water in the thermal fluids condensates to hot water and flows downward to heat the subjacent oil layer. At the same time, the non-condensate gas in the thermal fluids accumulates to the upper part of the oil layer and reduces heat loss. The analysis of heat loss shows that the heat loss rate gradually increases at first and then tends to be stable. Compared with conventional thermal fluid, the CSMTFS can more effectively reduce heat loss. The enthalpy value of supercritical multi-thermal fluid is significantly increased compared with that of multi-thermal fluid, which effectively solves the problem of insufficient heat carrying capacity of multi-thermal fluid. Overall, cyclic supercritical multi-thermal fluid stimulation can effectively solve the problems of conventional heavy oil thermal recovery technology in offshore heavy oil recovery; it is indeed a new improved-oil-recovery technique for offshore heavy oil. The findings of this study can help in better understanding the cyclic supercritical multi-thermal fluid stimulation process. This study is significant and helpful for application of CSMTFS technology in heavy oil recovery.

Keywords: heavy oil; physical simulation; supercritical multi-thermal fluid

1. Introduction

Heavy oil resources are abundant in the world, and their reserves are no less than conventional crude oil resources [1–6]. China is rich in heavy oil resources, and a large number of heavy oil resources are concentrated in the Bohai Bay. Due to the limitations of offshore platforms, conventional thermal recovery technologies (such as steam flooding, steam huff and puff, and conventional multi-thermal fluid) struggle to achieve good results at sea [7–11], which leads to a low degree of utilization of offshore heavy oil in China. Considering this situation, Zhou et al. [12] creatively combined supercritical water technology with heavy oil thermal recovery technology and proposed supercritical multi-component thermal fluid (SMTF) technology; that is, supercritical water reacts with diesel oil, heavy oil or oily sewage to generate supercritical multi-thermal fluid for offshore heavy oil thermal recovery. The main components of supercritical multi-thermal fluid include supercritical water, nitrogen and carbon dioxide, which are characterized by large heat carrying capacity, strong solubility and good diffusion.

Cyclic supercritical multi-thermal fluid stimulation (CSMTFS) is a specific form of supercritical multi-thermal fluid applied to heavy oil thermal recovery. The use of supercritical multi-thermal fluid for huff and puff development of offshore heavy oil can effectively overcome the problems existing in the offshore application of conventional thermal recovery technology, such as large heat loss, high dependence on fresh water and diesel oil, and high maintenance cost of complex pipelines. It is expected to improve the development effect of offshore heavy oil while reducing costs.

After supercritical multi-thermal fluid was proposed, many scholars paid attention to its generation process. Their studies [12–14] demonstrated through experiments that the generation process of supercritical multi-thermal fluid should be divided into two stages. First, supercritical water uses its own strong dissolution capacity to dissolve and disperse the organic substances in the reactants, and then, with the participation of oxygen, these organic substances are completely oxidized and decomposed in the supercritical water environment. These decomposed products, together with supercritical water, are called supercritical multi-component thermal fluids.

Subsequently, the interest of scholars was gradually attracted to the question of how to use supercritical multi-thermal fluid and its mechanism of heavy oil recovery; some research has focused on CMTFS. As the supercritical multi-thermal fluid also contains a large amount of nitrogen and carbon dioxide, the existing research on conventional thermal recovery technology also has certain reference value. Previous studies have confirmed that the addition of N_2, CO_2 and other non-condensable gases during the steam stimulation procedure can improve the steam stimulation recovery effect of heavy oil efficiently [15–18]. The results showed that, besides reducing viscosity by heating, multi-thermal fluid (MTF) also reduces viscosity by dissolving non-condensable gas (mainly CO_2) [19–23]. Moreover, the non-condensable gas (mainly) N_2 can also reduce the heat loss of the reservoir and maintain the reservoir pressure, thereby improving the recovery factor [24–29].

As the main component of supercritical multi- thermal fluid, supercritical water was once a great concern of scholars. These scholars successively carried out a series of research studies, which laid a foundation for the research of supercritical multi-thermal fluid. A deep heavy oil reservoir in northwest China took the lead in field testing of supercritical water huff and puff [30], and the production data showed that its recovery effect was improved compared with the conventional cyclic steam stimulation (CSS). On the basis of this study, scholars have carried out some research. Some studies discussed the process and characteristics of heavy oil recovery by supercritical water huff and puff technology through experimental research and theoretical analysis [31,32]. Some scholars focused on the supercritical hydrothermal cracking process of heavy oil, and discussed the reaction characteristics of the cracking process [33,34]. The results of these studies confirm that supercritical water can decompose macromolecular hydrocarbons in heavy oil into small molecule hydrocarbons and coke. With the gradual deepening of the research, the research on the mechanism of enhancing heavy oil recovery by supercritical water gradually emerged [35–38], among which [36] even started to conduct numerical simulation of this process.

However, as a new technology, although it has been gradually noticed by scholars, there is no research on the recovery effect and production characteristics of CSMTFS. In view of this situation, the L oil reservoir in Bohai was taken as the research object in this study, and a three-dimensional physical simulation experiment of CSMTFS was conducted by using a high temperature and high pressure model. During the experiment, the temperature and pressure distribution of the model and the production data of each cycle were obtained. With the results, heating effect, heat loss and production characteristics of CSMTFS were evaluated and analyzed. The issues discussed in this study are significant for a comprehensive understanding of the typical production characteristics of CSMTFS, and it can be used as a reference for CSMTFS oilfield practice.

2. Materials and Methods

2.1. Apparatus

The experimental apparatus consists of three main parts: injection–production system, reservoir model system and monitoring system, as shown in Figure 1. The injection–production system mainly includes distilled water, ISCO pumps, supercritical water generators (steam generators), non-condensable gas cylinders, pressurization pump, intermediate containers, hand pump and test tubes. The reservoir model system is a cube sand filling experimental device (Figure 2); the internal size of the model is 400 mm × 400 mm × 400 mm. The maximum working pressure of the model is 30 MPa and the maximum working temperature is 450 °C; the monitoring system mainly includes temperature sensors, pressure sensors, a data acquisition module and a computer.

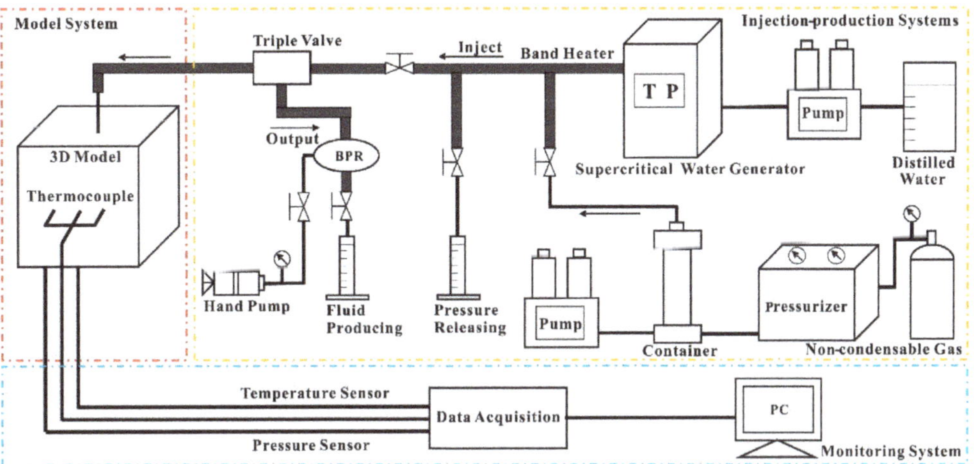

Figure 1. Flow diagram of three-dimensional physical simulation experiment.

Figure 2. Three-dimensional physical model.

As shown in Figure 3, the three-dimensional physical model is divided into 3 layers. A model well is placed vertically in the center of the oil layer, and the bottom of the well is 300 mm from the top of the model. Part of the wellbore in the middle sand layer is perforated. A total of 46 thermocouples are placed in the top layer and oil layer, and have the same distribution in both of them. Eight pressure probes are placed in the middle of the oil layer.

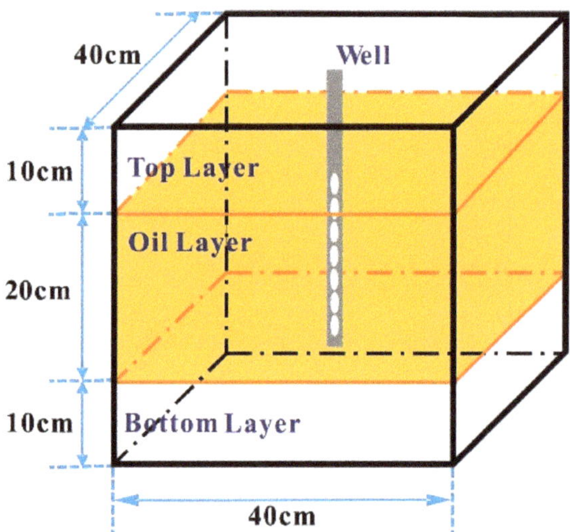

Figure 3. Schematic diagram of three-dimensional physical model.

2.2. Model Parameters

The similarity between the model and oilfield should be considered in three-dimensiona physical simulation. On the basis of experimental conditions, the experimental parameters of the three-dimensional physical simulation were determined by using the scaling criteria (Table 1) before the experiment.

Table 1. Scaling criteria.

Scaling Criteria	Physical Meaning	Modeling Parameters
$\pi_1 = \frac{Kg\Delta\rho}{v\mu_o}$	Ratio of gravity to viscous force	Permeability
$\pi_2 = \frac{\alpha_o}{vL}$	The ratio of heat conduction to convection	Steam injection rate
$\pi_3 = \frac{K\Delta p}{vL\mu_w}$	Ratio of driving force to viscous force	Differential pressure
$\pi_4 = \frac{vt}{L}$	Flow scale and space scale ratio	Time
$\pi_5 = \frac{\rho_g v_g}{\rho_s v_s}$	Ratio of mass fluidity of non-condensable gas to steam	Gas injection rate

Considering that the effective thickness of the oil layer in the L reservoir is about 40 m, while the thickness of the reservoir is designed to be 0.2 m based on the inner size of the experimental device, the geometrically similar ratio R can be obtained.

$$R = \frac{L_{mod}}{L_{prot}} = \frac{0.2}{40} = \frac{1}{200} \tag{1}$$

The crude oil used in this experiment was taken from the L reservoir and had been dehydrated and filtered before the experiment. The quartz sand used in the model has the same petro-physical properties as the L reservoir. Therefore, in the calculation process of experimental parameters, since the physical properties of the fluid and sand used in the experiment were the same as in the actual reservoir, the similarity ratio of the following parameters could be determined firstly: fluid density, fluid viscosity, oil saturation, porosity, initial reservoir temperature and initial reservoir pressure were all taken as 1. According to the scaling criteria, the experimental parameters of the three-dimensional physical model were determined, as shown in Table 2.

Table 2. Parameter conversion of oil reservoir prototype and scale model.

Parameter	Prototype	Model
Oil layer thickness, m	40	0.2
Porosity, %	40	40
Oil saturation, %	70	70
Permeability, mD	3500	700,000
Crude oil viscosity at 50 °C, mPa·s	4759	4759
Reservoir temperature, °C	50	50
Initial formation pressure, MPa	22	22
Steam injection rate (cold water equivalent), m^3/d	80	0.058
Cyclic cumulative steam injection volume (cold water equivalent), m^3	1600	0.0002
Cyclic injection increase, %	10	10
Soaking time, d	4	1/1440

Due to the limited experimental conditions of the laboratory, the steam injection rate directly converted according to the scaling criteria could not be satisfied, so it was necessary to determine a reasonable steam injection rate according to the experimental conditions. Considering the working conditions of the injection pump in the laboratory, the steam injection rate was designed to be 40 mL/min. At the same time, based on the cyclic cumulative steam injection volume and injection time, it could be calculated that 5 min of model injection was equivalent to 20 days of reservoir injection time, which meant that 1 year of reservoir recovery was equivalent to 82.5 min of model operation.

2.3. Procedures

Before the experiment begins, prepare the oil sample and sand required for the experiment, and then check the experimental apparatus to ensure that they can be operated normally. After that, connect the apparatus according to the experimental flow diagram (Figure 1).

(1) Load the model. Prepare the bottom layer with mud, and then fill the model with oil sands; during this process, place well, temperature and pressure measurement sensors in the predetermined position, as shown in Figure 4. After that, prepare the top layer.

Figure 4. Process of filling the three-dimensional physical model.

(2) Model initialization. After the loading process, check the hermeticity of the model. Then, move the model into the air bath. When the inner temperature of the model reaches 80 °C, saturate the crude oil into the model with a low flow rate to make the model pressure reach the predetermined pressure. Then, set the temperature of the air bath to 50 °C and age the model for no less than 48 h to simulate the initial state of the actual reservoir.

(3) Steam generator initialization. Before the SMTF is injected into the model, the steam generator should be heated and pressurized to make the injection temperature and injection pressure reach the predetermined values (390 °C, 22.1 MPa). At the same time, turn on the electric heating belt to preheat the steam injection pipeline.

(4) Experimental operation. During the experiment, the monitoring system records the temperature and pressure of the model, while the injection–production system injects SMTF into the model and collects the produced fluid in different cycles. The experiment ends when the predetermined maximum cycle is reached.

2.4. Methods

2.4.1. Analysis of Heating Effect

Due to the limitation of the experimental apparatus and the influence of system error, some experimental parameters could not meet the requirements of the scaling criteria. Therefore, individual parameters were adjusted according to the actual situation. Model and experimental parameters are shown in Tables 3 and 4.

Table 3. Parameters of three-dimensional physical model.

Parameter	Model
Oil layer thickness, m	0.2
Porosity, %	40
Oil saturation, %	70
Permeability, mD	10,000
Crude oil viscosity, mPa·s	4759
Initial formation temperature, °C	50
Initial formation pressure, MPa	22.1

Table 4. Parameters of supercritical multi-thermal fluids stimulation.

Parameters	1st Cycle	2nd Cycle	3rd Cycle	4th Cycle	5th Cycle	6th Cycle	7th Cycle	8th Cycle
Bottom hole temperature, °C	304.9	327.4	348.7	372.2	390.8	387.5	387.7	390.7
Steam injection rate (cold water equivalent), mL/min	40	40	40	40	40	40	40	40
Gas injection rate, mL/min	20	20	20	20	20	20	20	20
Injection time, min	4	4.5	5	5	5	5	5	5
Soaking time, min	1	1	1	1	1	1	1	1

The variation of temperature at different positions of the model could be obtained through the temperature sensors placed in the model; thus, the variation law of the heating range of CSMTFS could be intuitively analyzed.

2.4.2. Analysis of Production Characteristics

The change of oil production rate and cumulative oil production could be obtained by demulsifying the collected produced liquid, reading the data of oil and water, and calculating the instantaneous production and cumulative production of oil and water. In this way, the production law of CSMTFS could be analyzed intuitively.

2.4.3. Analysis of Heat Loss in Production Process

Reasonable simplification of the three-dimensional model was required to calculate the heat loss of CSMTFS. The calculation process had the following assumptions:

(1) During the experiment, only the heat loss at the top and bottom layers of the model was considered, while the heat loss at the side of the model was not considered. The reason was that the expansion of the heating area during the experiment did not reach the side boundary of the model, and the inner wall of the steel model was coated with thermal insulation material, which had extremely low thermal conductivity.

(2) During the experiment, the oil layer transferred heat to the top and bottom layers by heat conduction, without considering other forms of heat transfer.

(3) During the experiment, the outside temperature of the top and bottom layer was always the initial temperature of the reservoir (50 °C), while the inside temperature was measured by the temperature sensors.

Referring to the calculation method of heat loss in the existing literature [39], the heat flow through the top layer and bottom layer is:

$$\Phi = \frac{\lambda A}{\delta} \Delta T \tag{2}$$

where Φ is the heat flow through the top and bottom layer, J/min; ΔT is the temperature increment, °C; δ is the thickness of the top and bottom layer, cm; A is the heat conducting cross-section area, cm^2. λ is the thermal conductivity of the top and bottom layer; refer to the existing literature [40] for the thermal conductivity of the top and bottom layer of shale, the value is 2.57 J/(cm·min·°C).

On the plane perpendicular to the wellbore, taking the wellbore as the center, as shown in Figure 5, the heating area was divided into N heating rings with width dL = 1 cm. Each heating ring could be considered as an isothermal region, and its temperature was the arithmetic average temperature of the inner and outer boundaries of the heating ring.

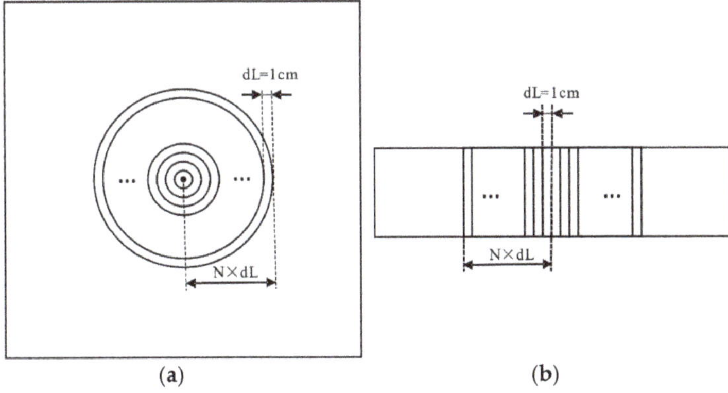

Figure 5. Heating area differential diagram. (**a**) overlooking; (**b**) side view.

At the end of injection process in each cycle, the sum of the heat fluxes at the top and bottom layers was equal to the heat loss rate, which can be expressed as:

$$Q_{tl} = \sum_{i=1}^{n} \Phi_{Ti} + \sum_{i=1}^{m} \Phi_{Bi} = \sum_{i=1}^{n} \frac{\lambda_c A_{Ti}}{\delta} \Delta T_{Ti} + \sum_{i=1}^{m} \frac{\lambda_r A_{Bi}}{\delta} \Delta T_{Bi} \tag{3}$$

The injection rate of heat was equal to the heat carried by the injection of supercritical multi-thermal fluid in unit time, and its expression is:

$$Q_i = i_{sc}(h_{sc} - h_g) \tag{4}$$

where Q_i is the heat injection rate, J/min; i_{sc} is the injection rate of supercritical multi-thermal fluid, g/min; h_{sc} is the enthalpy of supercritical multi-thermal fluid at bottom hole temperature, J/g; and h_g is the enthalpy of injected thermal fluid at utilization temperature (60 °C), J/g.

The ratio of heat loss rate to heat injection rate at a certain time was defined as the percentage of heat loss at that time.

3. Results

3.1. Heating Effect

The temperature distribution at the end of the injection process in each cycle is shown in Figures 6 and 7.

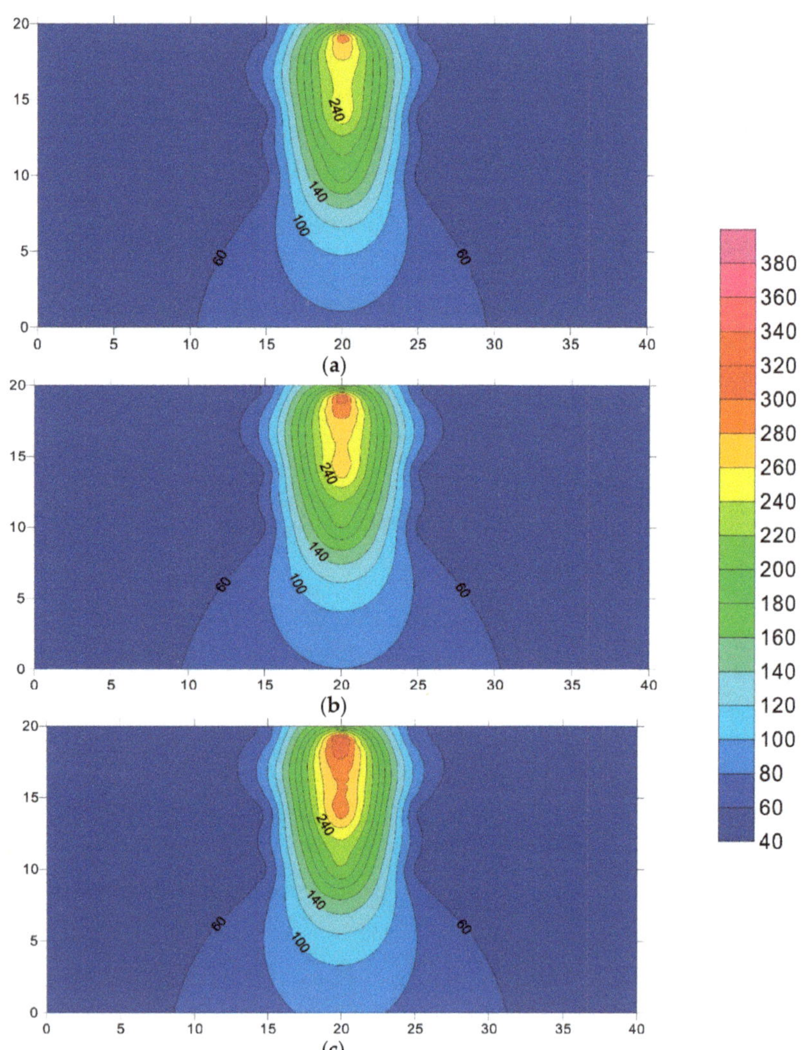

Figure 6. *Cont.*

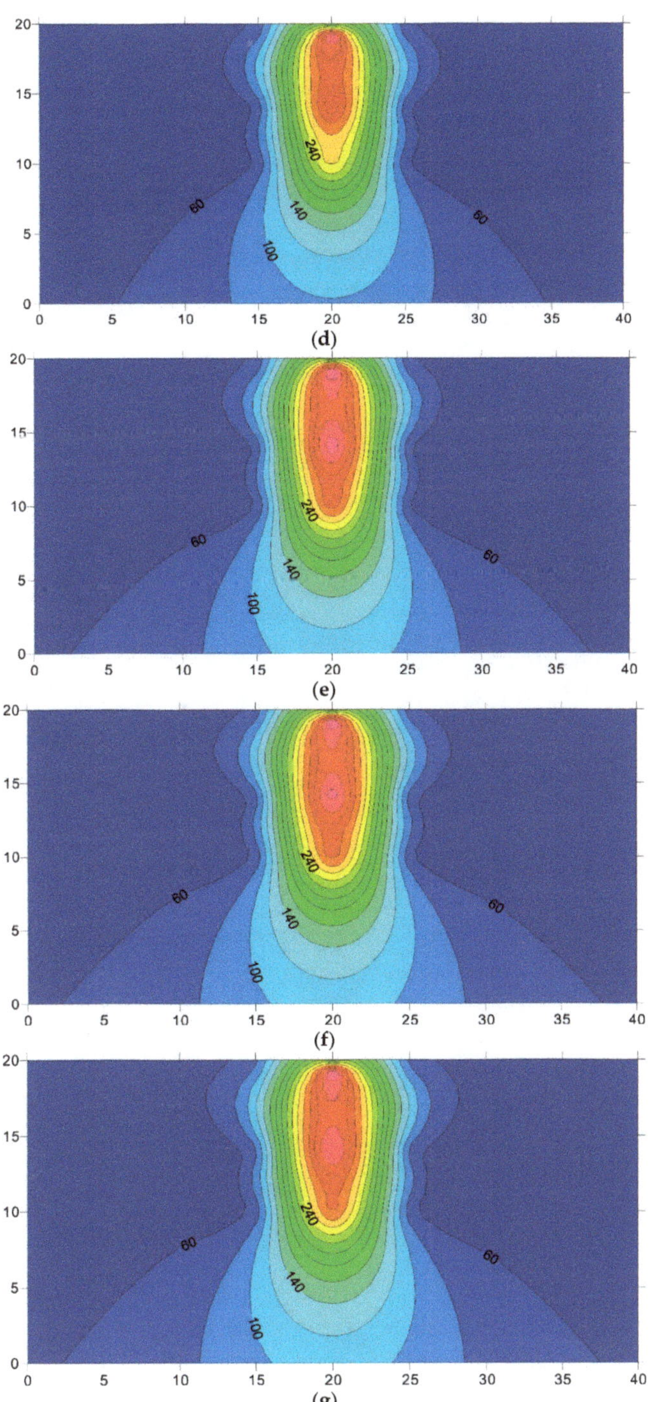

Figure 6. *Cont.*

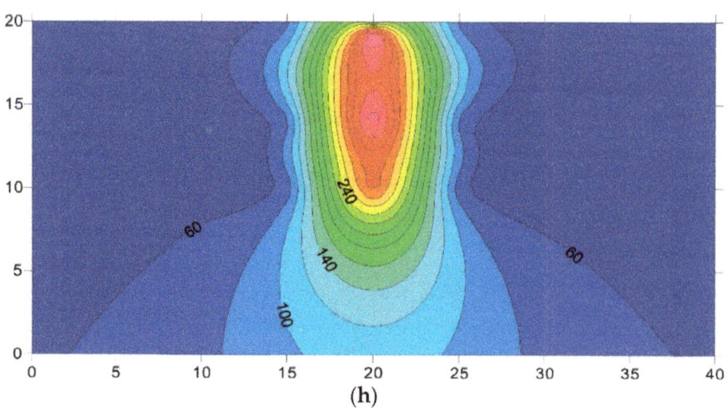

(h)

Figure 6. Vertical temperature profile of the oil layer in each cycle. (**a**) 1st cycle; (**b**) 2nd cycle; (**c**) 3rd cycle; (**d**) 4th cycle; (**e**) 5th cycle; (**f**) 6th cycle; (**g**) 7th cycle; (**h**) 8th cycle.

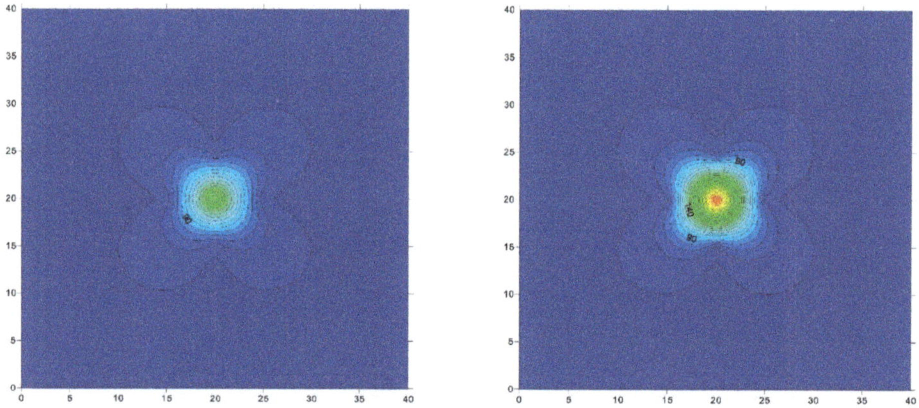

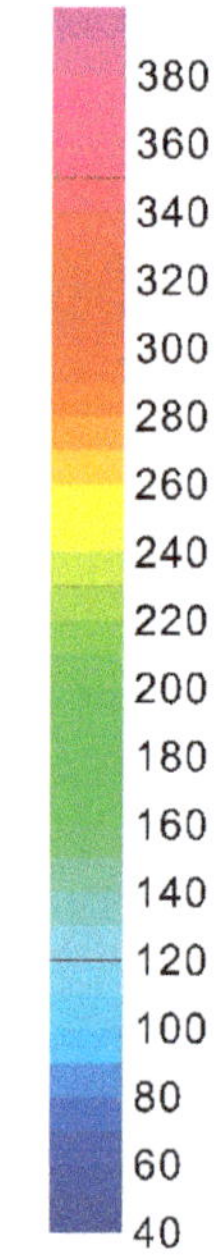

(**a**)

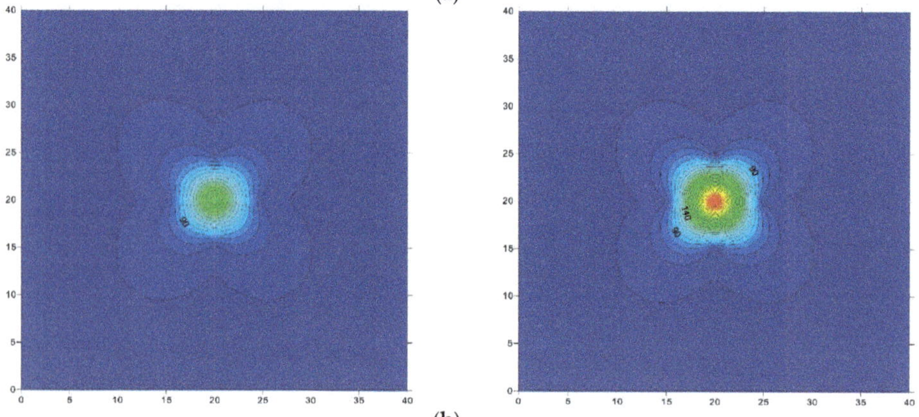

(**b**)

Figure 7. *Cont.*

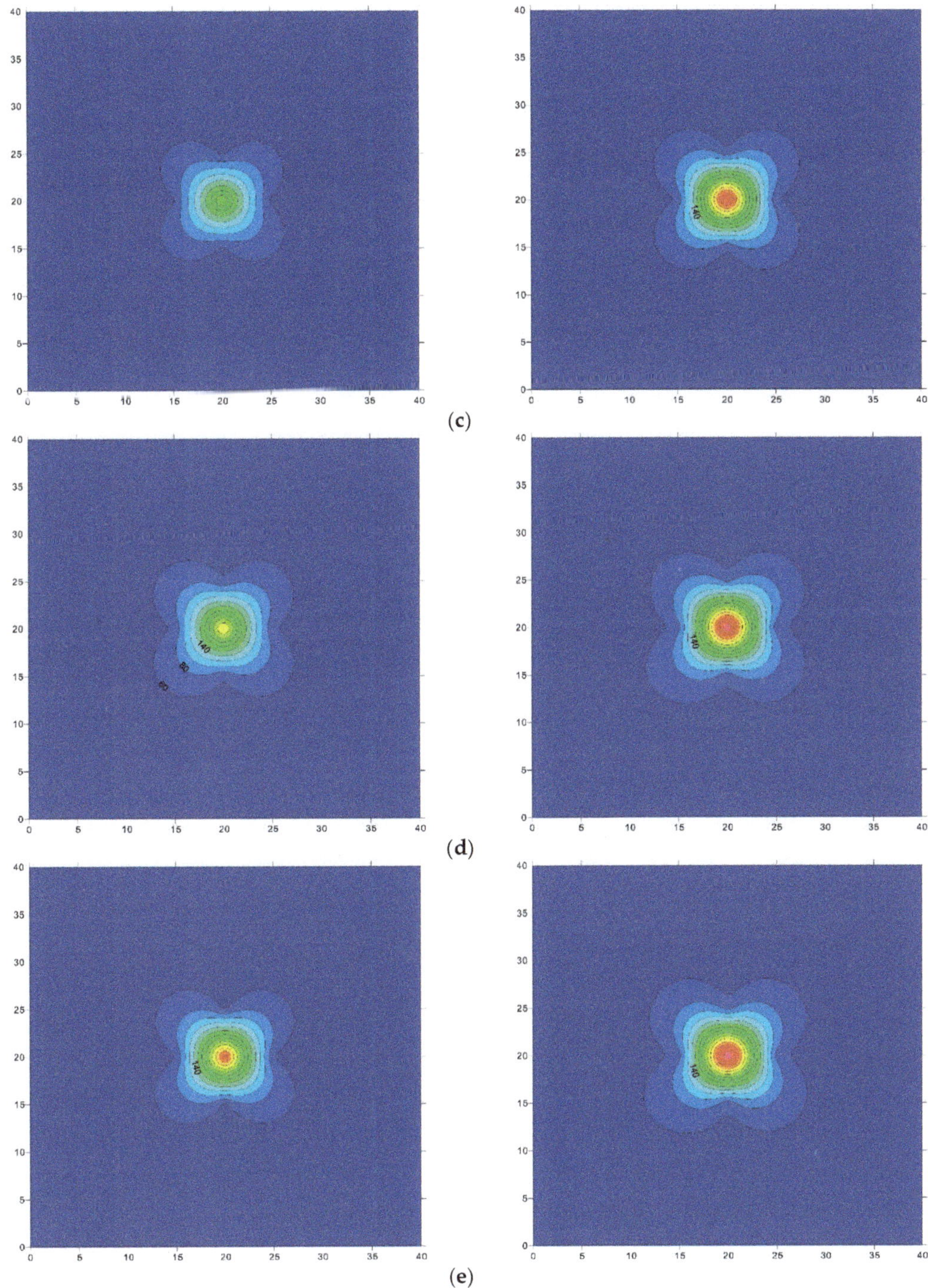

Figure 7. *Cont.*

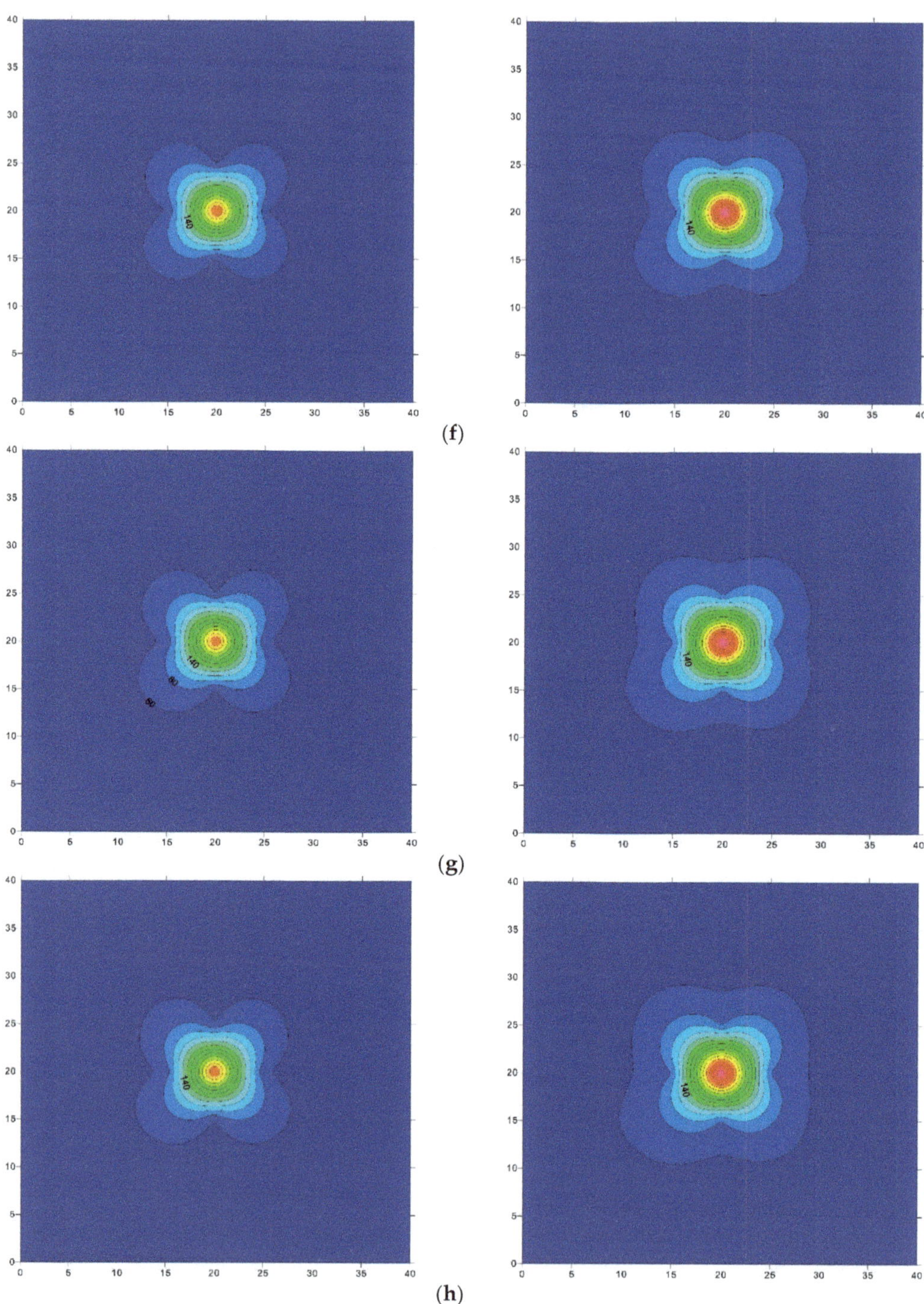

Figure 7. Horizontal temperature profile at the top (right) and middle (left) of the oil layer in each cycle. (**a**) 1st cycle; (**b**) 2nd cycle; (**c**) 3rd cycle; (**d**) 4th cycle; (**e**) 5th cycle; (**f**) 6th cycle; (**g**) 7th cycle; (**h**) 8th cycle.

The supercritical multi-thermal fluid is mainly composed of water, nitrogen and carbon dioxide. When the temperature and pressure of a fluid both exceed the critical temperature Tc and the critical pressure Pc, its state is called supercritical state. The critical points of water are 374.15 °C, 22.12 MPa, the critical points of nitrogen are −147.05 °C, 3.4 MPa, and the critical points of carbon dioxide are 31.3 °C, 7.39 MPa. As a result, as long as the water is in supercritical state, the thermal fluid is the supercritical multi-thermal fluid. Therefore, in the experimental process of this study, the critical temperature and pressure of water was used to determine whether the multi-thermal fluid was in supercritical state.

It can be seen in Figures 8 and 9 that, in the first four cycles, the temperatures at the top and middle of the oil layer were 304.9 °C, 327.4 °C, 348.7 °C and 372.2 °C, respectively. The temperature gradually increased, but the maximum temperature did not exceed 374.15 °C, so there was no supercritical multi-thermal fluid area in this stage. The reason was that when the experiment started, the temperature of the oil layer was not high enough. On the other hand, taking 60 °C as the temperature limit of the heating area in the oil layer, the heating area of the model had been effectively expanded in the first four cycles, and the radius of the heating area had increased from about 5 cm in the first cycle to about 7.5 cm in the fourth cycle. When the experiment proceeded to the fifth cycle, the heating area had further extended. More importantly, the temperature of the upper part of the oil layer had reached the maximum temperature of 390.8 °C, indicating that a supercritical multi-thermal fluid chamber began to shape in the oil layer at this time.

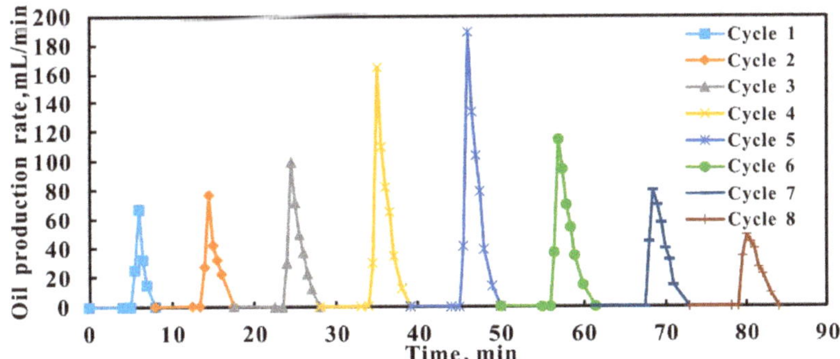

Figure 8. Oil production rate in each cycle.

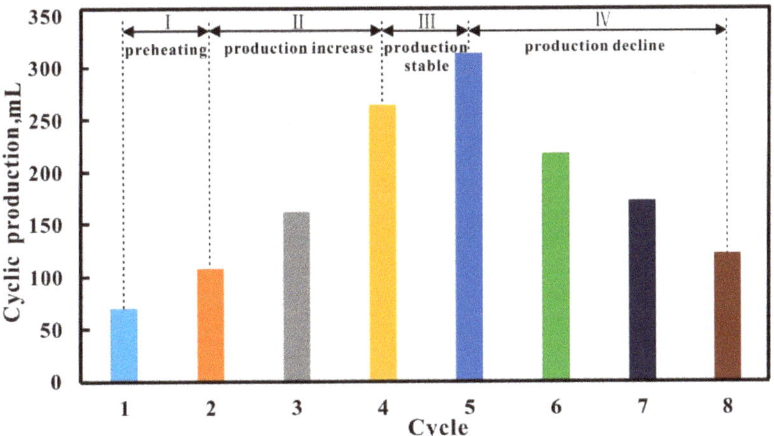

Figure 9. Production distribution in each cycle.

From the sixth to the eighth cycle, the temperature of the upper part of the oil layer was stable at about 390 °C, indicating that the thermal fluid heated the oil layer in supercritical state at first, but after its temperature dropped below the critical temperature, the thermal fluid heated the oil layer in hot water state.

3.2. Production Characteristic

The oil production rate, histogram and cumulative oil production of each cycle in the experiment are shown in Figures 8–10.

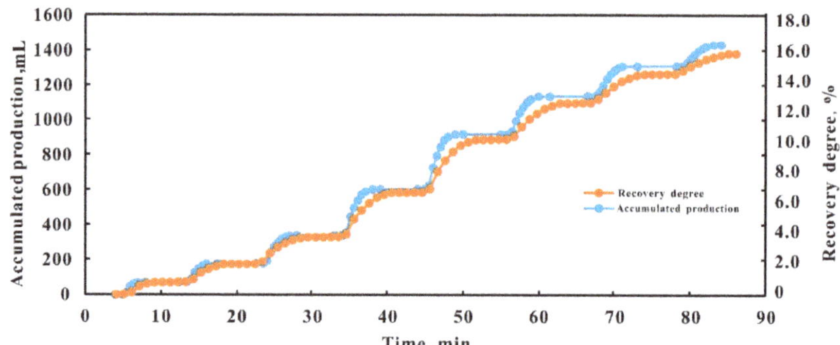

Figure 10. Cumulative oil production and recovery degree of the cyclic stimulation process.

Previous studies have shown that multi-thermal fluid stimulation has a high production and high oil production rate in the early stage, and its production continues to decline and the oil production rate decreases with the increase in stimulation cycles [8,36,41,42]. While in the process of CSMTFS, the production of each cycle increased to the peak first and then decreased gradually. On this basis, the production process was divided into four stages, namely, preheating stage, production increase stage, production stable stage and production decline stage, as shown in Figure 9.

Stage I is the preheating stage (first–second cycle). At this stage, the oil production rate was so low that the cumulative oil production increased slowly, and the increase in oil production rate was not obvious. In this stage, although the temperature of the injected supercritical multi-thermal fluid was very high, the reservoir temperature was equal to the original formation temperature, far lower than the temperature of the supercritical multi-thermal fluid. After the supercritical multi-thermal fluid was injected into the reservoir, it quickly dissipated heat and condensed to non-condensate gas and hot water, which could not effectively heat the reservoir. Moreover, the injection volume in the early stage was relatively small. These factors made the heating effect and diffusion ability of the thermal fluid both relatively poor, which led to a small heated area of the oil layer. The final manifestation was that the oil production rate at this stage was low, the cumulative oil production increased slowly and the oil production rate increased slightly.

Stage II is the production increase stage (second–fourth cycle). At this stage, the oil production rate increased rapidly and the cumulative oil production increased gradually. With the increase in stimulation cycle, the temperature of the oil layer gradually increased, and the supercritical multi-thermal fluid injected into the oil layer did not condense into hot water rapidly. In this process, the supercritical multi-thermal fluid chamber gradually formed, which could heat the oil layer more effectively. In this way, the range of heating area in the oil layer was obviously increased both vertically and horizontally, and the amount of crude oil that could flow was also significantly increased. Therefore, the oil production rate increases rapidly, and the cumulative oil production increases gradually at this stage.

Stage III is the production stable stage (fourth–fifth cycle). At this stage, the oil production rate reaches the peak gradually, and the cumulative oil production increases rapidly. During this process, the heating area of the supercritical multi-thermal fluid was basically unchanged, the shape of the formed supercritical multi-thermal fluid chamber was generally stable, and the supercritical multi-thermal fluid continuously and stably heated the oil layer. At this time, the oil saturation in the heating area was still at a high level, and the supply capacity of oil was strong. Therefore, the oil production rate was large and increased to the peak gradually, and the cumulative oil production increased rapidly.

Stage IV is the production decline stage (fifth–eighth cycle). At this stage, the oil production rate decreases gradually and the oil production increases slowly. The reason was that the injection temperature, injection rate and periodic injection volume were basically constant at the last stage of the CSMTFS process, while the expanding speed of the heating area gradually slowed down, and the stimulations of the first–eighth cycle made the oil saturation in the heating area of the oil layer gradually decrease to a low level. As a result, the heating area of the oil layer had insufficient oil supply capacity for the well, which led to a significant decline in oil production rate and a gradual slowdown in the increase in cumulative oil production at the later stage of the experiment.

3.3. Heat Loss in Production Process

As mentioned above, the variation of heat loss in each cycle of the CSMTFS process is calculated as shown in Figure 11.

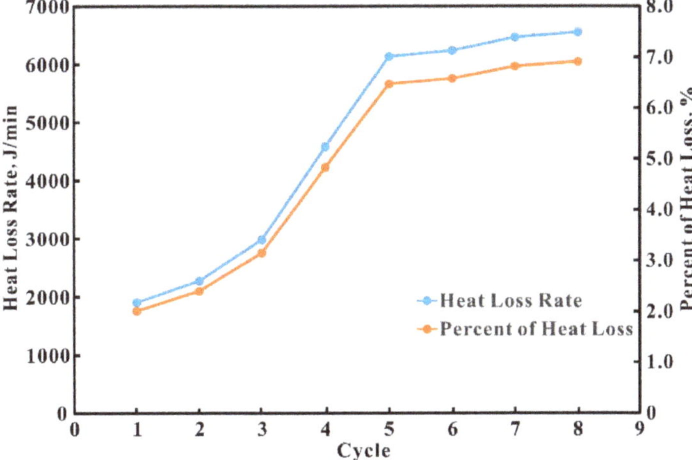

Figure 11. Heat loss rate and percentage in each cycle.

In the preheating stage (first–second cycle), most of the supercritical multi-thermal fluid was rapidly condensed into hot water and the heated area in the oil layer was very small, which led to the area of heat loss on the top and bottom layer being small as well. As a result, heat loss rate and heat loss percentage were at a low level. In the production increase stage (second–fourth cycle), the oil layer was gradually and effectively heated, and the heating range was continuously expanded, so the area of heat loss in the top and bottom layer was gradually increased. Based on this, the heat injection rate, heat loss rate and heat loss percentage increased continuously, and the heat loss percentage increased to the maximum in the fourth cycle. In the production stable stage (fourth–fifth cycle), the heating range and supercritical multi-thermal fluid chamber were expanded to the maximum and remained stable. The area of heat loss in the top and bottom layers also expanded rapidly; thus, the heat injection rate and heat loss rate increased rapidly, which made the percentage of heat loss increase to the maximum in the fourth cycle and then decrease slightly. In the production decline stage (fifth–eighth cycle), the heating range and supercritical multi-thermal fluid chamber remained stable, and the area of heat loss generated in the top and bottom layer was also stable. Therefore, the heat injection rate was basically stable, while the heat loss rate and percentage of heat loss increased slightly.

Based on the temperature distribution at different production stages in the process of multi-thermal fluid flooding [27], the heat loss at different production stages could be obtained by calculation. The calculation method of heat loss was similar to that of CSMTFS. Based on the parameters in reference [43], the calculated heat injection rate of multi-thermal fluid flooding is 14,380.5 J/min, and the calculation results of heat loss at each production stage are shown in Table 5.

Table 5. Heat loss at different production stages in multi-thermal fluid flooding.

Production Stage	Heat Loss Rate, J/min	Percentage of Heat Loss, %
Rapid effect stage	323.7	2.25
Rapid decline stage	742.3	5.16
Slow decline stage	1343.4	9.34

The calculation results showed that the heat loss rate and percentage of heat loss increased rapidly with the progress of the production stage in the multi-thermal fluid flooding. After the production entered the slow decline stage, the heat loss percentage reached the maximum value of 9.34%, which was higher than the peak value of 7.12% in the process of CSMTFS. The temperature of the supercritical multi-thermal fluid was much higher than that of the thermal fluid in the literature [21], but it can be seen from the results in Table 5 and Figure 11 that the heat loss during the CSMTFS process was smaller, which indicates that CSMTFS can more effectively reduce heat loss.

4. Discussion

During all stimulation cycles, there was always a low temperature zone on the top of the oil layer at the end of the injection process. The reason was that the non-condensable gas (mainly nitrogen) existing in the supercritical multi-thermal fluid accumulated to the upper part of the oil layer under the action of gravity differentiation and formed a thermal insulation layer. Before the stimulation experiment entered the last stage (fifth–eighth cycle), there was no supercritical multi-thermal fluid chamber in the model, which indicated that the oil layer was heated by thermal fluid in a hot water state after the thermal fluid was injected into the model. After the experiment entered the late stage, a small supercritical multi-thermal fluid chamber (about 2 cm in radius) was formed in the upper part of the model after the supercritical multi-thermal fluid was injected into the model. Outside the supercritical multi-thermal fluid chamber, the thermal fluid still existed in the hot water state and flowed downward under the action of gravity to heat the lower part of the model. In the cyclic stimulation process, although the supercritical multi-thermal fluid chamber was relatively small, the heat carrying capacity of the supercritical multi-thermal fluid was much higher than that of multi-thermal fluid, which effectively solved the problem of insufficient heat carrying capacity of multi-thermal fluid stimulation. Previous studies have shown that the injection temperature of conventional multi-thermal fluid stimulation is generally below 300 °C, and the optimized temperature under field conditions is mostly 240 °C; the injection pressure is 10 MPa [9,44,45]. Under this condition, the main heat carrier of the multi-thermal fluid is steam and its enthalpy is only 1038.32 kJ/kg. By contrast, in this experiment, the injection pressure was 22.1 MPa and the minimum injection temperature was 304.9 °C, while the maximum injection temperature was 390.8 °C; the enthalpy of steam under this condition is 1358.68–2640.39 kJ/kg, which is 30.85–154.29% higher than that of conventional multi-thermal fluid.

In the cyclic stimulation process, the production stable stage was the main oil production stage, which accounted for 40.32% of the total oil production. The production decline stage was the subsidiary oil production stage, which accounted for 35.81% of the total oil production. In contrast, the sum of oil production of the preheating stage and production increase stage was lower, accounting for only 23.87% of the total oil production.

The reason was that in the early stage of the experiment, the heating effect on the oil layer was relatively poor and the production was relatively low. Taking into account the effect of cyclic injection volume on production in the process of multi-thermal fluid injection [41,42], in the production process, the cyclic injection volume can be appropriately increased in the early stage of cyclic supercritical multi-thermal fluid stimulation to shorten the time required for the preheating stage and thus improve production.

The production process in a single cycle of cyclic supercritical multi-thermal fluid stimulation was similar to that of multi-thermal fluid stimulation [46], which could be divided into three stages, namely, the drainage stage, production increase stage and production decline stage. The oil production rate increased from a low level to the peak and then decreased gradually. The reason was that after the supercritical multi-thermal fluid was injected into the model, the water saturation and temperature in the area near the wellbore were very high, and there was a drainage period in the early stage of well opening after soaking. After the drainage stage, the temperature of the heated area was still high and its liquid supply capacity was strong, so the oil production rate was large. With the increase in production time, as the heat of the oil layer was lost from the top and bottom layers and carried out by the output liquid, the temperature of the heated area was gradually reduced, and the pressure of the oil layer was also gradually reduced, so the oil production rate began to decline rapidly. In contrast to the high-production and low-production periods in the production experiment of multi-thermal fluid stimulation [8,36], there was no production stable period in the laboratory experiment, because the volume of the model was small and the amount of stored elastic potential energy was small.

This study shows that CSMTFS has the characteristics of high heat carrying capacity and minor heat loss compared with conventional stimulation methods (CMTFS and CSS), so it can efficiently enhance the heating effect of huff and puff operation on the formation and improve thermal efficiency. In addition, the injection of multi-thermal fluids and steam has the problems of high fuel consumption and high water treatment requirements, which leads to high production cost, and this problem is particularly prominent in the recovery of offshore heavy oil reservoirs. The generation of supercritical multi-thermal fluid can directly use oily wastewater, which greatly reduces the cost of fuel and water treatment, and thus improves the economic benefit of the oilfield. At the same time, due to the complex steam generation process of offshore platforms, there are many steam generation and injection pipelines, and pipeline leakage and corrosion problems are prominent. In contrast, the generation process of supercritical multi-thermal fluid is relatively simple, which can significantly reduce the cost of pipeline maintenance. Of course, due to the high temperature and high pressure of supercritical multi- thermal fluid, it puts forward higher requirements for the temperature resistance and pressure resistance of various pipelines, which can be considered as a limitation of this technology. In general, supercritical multi-thermal fluid stimulation meets the demand of "cost reduction and efficiency increase" in the current oil and gas field development process, and can be used as an important alternative technology for offshore heavy oil recovery.

It is worth pointing out that, since the recovery of crude oil is only about 16% through nine cycles of huff and puff in this study compared with conventional heavy oil thermal recovery technology, it may seem that cyclic supercritical multi-thermal fluid stimulation does not show obvious advantages in enhancing heavy oil recovery. However, it should be noted that the oil well in this study only penetrates half of the thickness of the oil layer. If all the oil layers are drilled through and the process parameters are reasonably optimized, the degree of crude oil recovery through cyclic supercritical multi-thermal fluid stimulation will most likely exceed 25%, which is superior to the development effect achieved by other conventional thermal fluid huff and puff. The technique of cyclic supercritical multi-thermal fluid stimulation is the application of supercritical multi-thermal fluid in heavy oil recovery. Although the supercritical multi-thermal fluid was first proposed for the problems of offshore heavy oil recovery, it also can be applied to onshore heavy oil reservoirs, especially for onshore deep heavy oil reservoirs. In conclusion, cyclic supercritical multi-thermal fluid stimulation can indeed overcome the problems of conventional heavy oil thermal recovery technology in offshore heavy oil development, and is expected to achieve better development results while reducing costs.

5. Conclusions

Through this study on the process of CSMTFS, the following conclusions can be obtained:

(1) The whole development process of CSMTFS can be divided into four stages, namely, the preheating stage, production increase stage, production stable stage and production decline stage. The production stable stage was the main oil production stage, while the production decline stage was the secondary oil production stage. The sum of oil production in these two stages accounted for 76.13% of the total oil production.

(2) The temperature and heating area of the oil layer increased first and then tended to be flat. There was no supercritical multi-thermal fluid chamber in the early and middle stages of the CSMTFS process, and only a small range of supercritical multi-thermal fluid chamber was formed in the last stage of the CSMTFS process. The enthalpy value of supercritical multi-thermal fluid was significantly increased compared with that of multi-thermal fluid, which effectively solved the problem of insufficient heat carrying capacity of multi-thermal fluid.

(3) In the process of cyclic supercritical multi-thermal fluid stimulation, the percentage of heat loss increases first and then tends to be stable; the maximum is 7.12%. Compared with conventional heavy oil thermal recovery technology, the percentage of heat loss in the process of cyclic supercritical multi-thermal fluid stimulation is lower.

Cyclic supercritical multi-thermal fluid stimulation has effectively solved the problems of conventional heavy oil thermal recovery technology in offshore heavy oil recovery, and has significantly improved the development effect, cost, heat carrying capacity, heat loss, etc. It is indeed a new improved-oil-recovery technique for offshore heavy oil.

Author Contributions: J.T., W.Y., Z.Q. and Y.Y. conceived and designed the experiments; S.H., M.D. and J.T. conducted the experiments and analyzed the data; J.T. wrote the paper. All authors have read and agreed to the published version of the manuscript.

Funding: This research was funded by National Natural Science Foundation of China (grant number 52004048), Natural Science Foundation of Chongqing Municipality, China (grant number cstc2020jcyj msxmX0856) and Science and Technology Research Project of the Chongqing Municipal Education Commission of China (grant number KJQN201901542).

Data Availability Statement: The data that support the findings of this study are available from the corresponding author.

Conflicts of Interest: The authors declare no conflict of interest.

References

1. Zhao, S.; Pu, W.F.; Su, L.; Shang, C.; Song, Y.; Li, W.; He, H.Z.; Liu, Y.G.; Liu, Z.Z. Properties, combustion behavior, and kinetic triplets of coke produced by low-temperature oxidation and pyrolysis: Implications for heavy oil in-situ combustion. *Pet. Sci.* **2021**, *18*, 1483–1491.
2. Wei, W.; Ian, D.G. Lag times in toe-to-heel air injection (THAI) operations explain underlying heavy oil production mechanisms. *Pet. Sci.* **2022**, *19*, 1165–1173.
3. Sun, X.F.; Song, Z.Y.; Cai, L.F.; Zhang, Y.Y.; Li, P. Phase behavior of heavy oil–solvent mixture systems under reservoir conditions. *Pet. Sci.* **2020**, *17*, 1683–1698. [CrossRef]
4. Cao, Y.; Chen, W.; Wang, T.; Yuan, Y.N. Thermally enhanced shale gas recovery: Microstructure characteristics of combusted shale. *Pet. Sci.* **2020**, *17*, 1056–1066.
5. Zhu, C.F.; Guo, W.; Wang, Y.P.; Li, Y.J.; Gong, H.J.; Xu, L.; Dong, M.Z. Experimental study of enhanced oil recovery by CO_2 huff-n-puff in shales and tight sandstones with fractures. *Pet. Sci.* **2021**, *18*, 852–869.
6. Zhu, D.Y. *Oilfield Chemistry*, 1st ed.; Petroleum Industry Press: Beijing, China, 2018.
7. Liu, D.; Hu, T.H.; Pan, G.M.; Wu, J.T.; Zhang, J.T. Comparison of Production Results between Multiple Thermal Fluid Huff and Puff and Steam Huff and Puff in Offshore Application. *Spec. Oil Gas Reserv.* **2015**, *22*, 118–120.
8. Huang, Y.H.; Liu, D.; Luo, Y.K. Research on multiple thermal fluid stimulation for offshore heavy oil production. *Spec. Oil Gas Reserv.* **2013**, *20*, 84–86.

9. Chen, J.B. Multiple Thermal Fluid Huff-Puff in Offshore Deep Thin Heavy Oil Reservoir. *Spec. Oil Gas Reserv.* **2016**, *23*, 97–100.
10. Sun, F.R.; Yao, Y.D.; Li, X.F.; Zhang, Y.; Ding, G.Y. Evaluation of Heating Effect on the Horizontal Well in 35-2Bohai Oilfield with Multiple Thermal Fluid Stimulation. *J. Beijing Inst. Petrochem. Technol.* **2017**, *25*, 5–8.
11. Wang, X.G.; Qiao, W.J. Research and Practice of Application Method for Thermal Recovery Technology for Offshore Multiple Thermal Fluid. *Sino-Glob. Energy* **2014**, *19*, 37–41.
12. Zhou, S.W.; Guo, L.J.; Li, Q.P. Multi-source Multi- thermal Fluid Generation System and Method. CN Patent CN106640007A, 10 May 2017.
13. Xu, J.L.; Kou, J.J.; Guo, L.J. Experimental study on oil-containing wastewater gasification in supercritical water in a continuous system. *Int. J. Hydrog. Energy* **2019**, *44*, 15871–15881. [CrossRef]
14. Sun, X.F.; Cai, J.M.; Li, X.Y. Experimental investigation of a novel method for heavy oil recovery using supercritical multi thermal fluid flooding. *Appl. Therm. Eng.* **2020**, *185*, 116–130.
15. Wang, C.J.; Liu, H.Q.; Wang, J.; Wu, Z.; Wang, L. Three-dimensional physical simulation experiment study on carbon dioxide and dissolver assisted horizontal well steam stimulation in super heavy oil reservoirs. *J. Pet. Explor. Prod. Technol.* **2016**, *6*, 825–834.
16. Wu, R.N.; Wei, B.; Zou, P.; Zhang, X.; Shang, J.; Gao, K. Effect of Supercritical CO_2 on the Physical Properties of Conventional Heavy Oil and Extra-heavy Oil. *Oilfield Chem.* **2018**, *35*, 64–70.
17. Zhang, X.L. *Study on the Mining Mechanism and Adaptability of Multi Thermal Fluid in Heavy Oil Reservoir*; China University of Petroleum (East China): Qingdao, China, 2017.
18. Zhang, F.Y.; Xu, W.K.; Wu, T.T.; Ge, T.T.; Wang, H.J.; Wu, C.X.; Wang, D.W. Research on the mechanism of multi-thermal fluids on enhanced oil recovery and reservoir adaptability. *Pet. Geol. Recovery Effic.* **2014**, *21*, 75–78.
19. Gu, H.; Cheng, L.S.; Zhang, X.L.; Zhao, Y.Q.; Peng, P. A Study on Nitrogen-injection Technology by Horizontal Well in Low-permeability Shallow-thin Heavy Oil Reservoir Science. *Technol. Eng.* **2013**, *13*, 9494–9497.
20. Du, Y.; Wang, Y.; Jiang, P.; Ge, J.; Zhang, G. Mechanism and Feasibility Study of Nitrogen Assisted Periodic Steam Stimulation for Ultra-Heavy Oil Reservoir. In Proceedings of the SPE Enhanced Oil Recovery Conference, Kuala Lumpur, Malaysia, 2–4 July 2013.
21. Moussa, T.M.; Patil, S.; Mahmoud, M.A. Performance and Economic Analysis of a Novel Heavy Oil Recovery Process Using In-Situ Steam and Nitrogen Generated by Thermochemicals. In Proceedings of the SPE Kingdom of Saudi Arabia Annual Technical Symposium and Exhibition, Dammam, Saudi Arabia, 23–26 April 2018.
22. Liu, D.; Li, Y.P.; Zhang, F.Y.; Zhang, L.; Zhang, C.Q. Reservoir applicability of steam stimulation supplemented by flue gas. *China Offshore Oil Gas* **2012**, *24*, 62–66.
23. Sun, F.R.; Yao, Y.D.; Li, X.F.; Zhao, L.; Ding, G.; Zhang, X. The mass and heat transfer characteristics of superheated steam coupled with non-condensing gases in perforated horizontal wellbores. *J. Pet. Sci. Eng.* **2017**, *156*, 460–467.
24. Lv, S.Y.; Li, Y.H.; Li, H.B.; Long, X.M. Predicting model of the physical property parameter of super-critical water injection wellbore in horizontal well of heavy oil reservoirs. *Pet. Geol. Oilfield Dev. Daqing* **2021**, *40*, 54–62.
25. Cheng, W.L.; Han, B.B. Wellbore heat transfer model of multiple thermal fluid based on real gas state equation. *Acta Pet. Sin.* **2015**, *36*, 1402–1410.
26. Jia, X.F.; Ma, K.Q.; Liu, Y.X.; Liu, B.; Zhang, J.; Li, Y. Enhance Heavy Oil Recovery by In-Situ Carbon Dioxide Generation and Application in China Offshore Oilfield. In Proceedings of the SPE Enhanced Oil Recovery Conference, Kuala Lumpur, Malaysia, 2–4 July 2013.
27. Liu, D.; Su, Y.C.; Chen, J.B.; Zhang, C.Q.; Pan, G.M. A 3-D Physical Simulation Experiment and Numerical Test on Multi-thermal Fluids Flooding After Huff and Puff. *J. Southwest Pet. Univ.* **2019**, *41*, 140–149.
28. Liu, Y.G.; Yang, H.L.; Zhao, L.C.; Sun, Y.; Cui, G.; Zhao, M.; Zhong, L. Improve Offshore Heavy Oil Recovery by Compound Stimulation Technology Involved Thermal, Gas and Chemical Methods. In Proceedings of the Offshore Technology Conference, Houston, TX, USA, 3–6 May 2010.
29. Zhong, L.G.; Yu, D.; Yang, H.L.; Sun, Y.; Wang, G.; Zheng, J. Feasibility Study on Producing Heavy Oil by Gas and Electrical Heating Assisted Gravity Drainage. In Proceedings of the Offshore Technology Conference, Houston, TX, USA, 2–5 May 2011.
30. Yin, Q.G.; Huo, H.B.; Hang, P.; Wang, C.; Wu, D.; Luo, Z. Effective Use of Technology Research and Application of Supercritical Steam on Deep Super-heavy Oil in Lukeqin Oilfield. *Oilfield Chem.* **2017**, *34*, 635–641.
31. Wang, R. *Research on Supercritical Cyclic Steam Stimulation of Ultra-Deep Super Heavy Oil in East Area of Lukeqin*; University of Petroleum (East China): Qingdao, China, 2015.
32. Zhao, Q.Y.; Guo, L.J.; Huang, Z.J.; Chen, L.; Jin, H.; Wang, Y. Experimental Investigation on Enhanced Oil Recovery of Extra Heavy Oil by Supercritical Water Flooding. *Energy Fuels* **2018**, *32*, 1685–1692.
33. Zhang, F.Y.; Pang, Z.X.; Wu, T.T.; Geng, Z.G.; Liao, H. Extra-Heavy Oil Hydrothermal Pyrolysis Experiment with Supercritical Steam. *Spec. Oil Gas Reserv.* **2019**, *26*, 130–135.
34. Zhao, Q.Y.; Guo, L.J.; Wang, Y.H.; Jin, H.; Chen, L.; Huang, Z. Enhanced Oil Recovery and in Situ Upgrading of Heavy Oil by Supercritical Water Injection. *Energy Fuels* **2019**, *34*, 360–367.
35. Zhao, Q.Y.; Guo, L.J.; Wang, Y.C.; Huang, Z.; Chen, L.; Jin, H. Thermophysical Characteristics of Enhanced Extra-heavy Oil Recovery by Supercritical Water Flooding. *J. Eng. Thermophys.* **2020**, *41*, 635–642.
36. Wang, S.T.; Zhang, F.Y.; Liu, D.; Zhu, Q.; Ge, T.T. Physical—Numerical Simulation Integration of Super—Critical Steam—Flooding in Offshore Extra—Heavy Oil Reservoir. *Spec. Oil Gas Reserv.* **2020**, *27*, 138–144.

37. Zhang, F.Y.; Wang, S.T.; Liu, D.; Gao, Z.; Zhu, Q. Experimental study on the EOR mechanisms of developing extra heavy oil by supercritical steam. *Oil Drill. Prod. Technol.* **2020**, *42*, 242–246.
38. Han, S. *Mechanism and Application of Supercritical Cyclic Steam for Heavy Oil Reservoirs*; Northeast Petroleum University: Daqing, China, 2020.
39. Tian, J.; Liu, H.Q.; Pang, Z.X.; Zhao, W.; Gao, Z.N. Experiment of 3D physical simulation on dual horizontal well SAGD under high pressure condition. *Acta Pet. Sin.* **2017**, *04*, 95–102.
40. Liu, W.Z. *Thermal Recovery of Heavy Oil by Steam Injection*; Petroleum Industry Press: Beijing, China, 1997.
41. Yang, B.; Li, J.S.; Zhang, X.S.; Jiang, J.; Gong, R.X. Study on horizontal well multi-component thermal fuid stimulation high efficiency recovery technology for heavy oil reservoir. *Pet. Geol. Recovery Effic.* **2014**, *21*, 41–44.
42. Liu, Y.; Wang, C.X.; Li, X.H.; Xu, H. Development mechanism of multivariate thermal fluid of deep super-heavy oil. *Fault-Block Oil Gas Field* **2019**, *26*, 98–103.
43. Liang, W.; Zhao, X.H.; Zhang, Z.J.; Liu, H.; Qiao, Z. Action mechanism and application of multiple-thermal fluids to improve heavy oil recovery. *Pet. Geol. Eng.* **2014**, *28*, 115–117.
44. Liang, D.; Feng, G.Z.; Zeng, X.L.; Fang, M.J.; He, C.B. Numerical simulation study on multiple thermal fluid throughput of Bohai heavy oilfield. *Pet. Geol. Eng.* **2013**, *27*, 130–132.
45. Yang, B.; Li, J.S.; Qi, C.X.; Shi, H.L.; Zhu, W.X. Research on optimized multiple thermal fluids stimulation of offshore heavy oil reservoirs. *Pet. Geol. Eng.* **2012**, *26*, 54–56.
46. Zhang, W.; Sun, Y.T.; Lin, T.; Ma, Z.H.; Sun, Y.B. Experimental study on mechanisms of the Multi-fluid thermal recovery on offshore heavy oil. *Petrochem. Ind. Appl.* **2013**, *32*, 34–36.

MDPI

Article

A Study on Generation and Feasibility of Supercritical Multi-Thermal Fluid

Xiaoxu Tang [1,2], Zhao Hua [1,3], Jian Zhang [1,3], Qiang Fu [3] and Jie Tian [4,*]

1 State Key Laboratory of Offshore Oil Exploitation, Beijing 100028, China
2 Tianjin Branch of CNOOC China Ltd., Tianjin 300452, China
3 CNOOC Research Institute Co., Ltd., Beijing 100028, China
4 School of Petroleum and Natural Gas Engineering, Chongqing University of Science & Technology, Chongqing 401331, China
* Correspondence: tianjie@cqust.edu.cn

Abstract: Supercritical multi-thermal fluid is an emerging and efficient heat carrier for thermal recovery of heavy oil, but the generation of supercritical multi-thermal fluid and its feasibility in thermal recovery are rarely discussed. In this paper, generation and flooding experiments of supercritical multi-thermal fluid were carried out, respectively, for the generation and feasibility of supercritical multi-thermal fluid. During the experiment, the temperature and pressure in the reactor and sand-pack were monitored and recorded, the fluid generated by the reaction was analyzed by chromatography, and enthalpy of the reaction product and displacement efficiency were calculated, respectively. The experimental results showed that the change in temperature and pressure in the reactor could be roughly divided into three stages in the generation process of supercritical multi-thermal fluid. The higher the proportion of oil in the reactant, the higher the maximum temperature in the reactor. When the proportion of oil and water in the reactant was constant, the temperature rise in the reactor was basically the same under different initial temperature and pressure conditions. Compared with the initial temperature and pressure, the oil–water ratio of the reactants had a significant effect on the generated supercritical multi-thermal fluid. The higher the proportion of oil, the more gas that was generated in the supercritical multi-thermal fluid, and the lower the specific enthalpy of the thermal fluid. Under the same proportion of oil and water, the gas–water mass ratio of the supercritical multi-thermal fluid generated by the reaction of crude oil was lower, and the specific enthalpy was higher. Through this study, it was found that supercritical multi-thermal fluid with a low gas–water mass ratio had higher oil displacement efficiency, higher early oil recovery rate, a larger supercritical area formed in the oil layer, and later channeling. The results of this study show that the optimal gas–water mass ratio of supercritical multi-thermal fluid was about 1, under which the oil displacement efficiency and supercritical area in the oil layer reached the maximum. Correspondingly, the optimal proportion of oil in the reactant when generating supercritical multi-component thermal fluid was about 10%. In oilfield applications, because the gas–water ratio in supercritical multi-component thermal fluid has a significant impact on oil displacement efficiency, the optimization of supercritical multi-thermal fluid should not only consider the generation process but also consider the oil displacement effect of the thermal fluid. The findings of this study could improve our understanding of the characteristics of generating supercritical multi-thermal fluid and the feasibility of supercritical multi-thermal fluid generated under different conditions in the oil displacement process. This research is of great significance for field applications of supercritical multi-thermal fluid.

Keywords: heavy oil; supercritical multi-thermal fluid; thermal recovery

Citation: Tang, X.; Hua, Z.; Zhang, J.; Fu, Q.; Tian, J. A Study on Generation and Feasibility of Supercritical Multi-Thermal Fluid. *Energies* **2022**, *15*, 8027. https://doi.org/10.3390/en15218027

Academic Editor: Daoyi Zhu

Received: 30 August 2022
Accepted: 24 October 2022
Published: 28 October 2022

1. Introduction

The world is extremely rich in heavy oil resources, and its geological reserves far exceed conventional crude oil. The thermal recovery method is the most effective method to recover heavy oil at present, mainly including cyclic steam stimulation, steam flooding, SAGD, in-situ combustion, hot water flooding, etc. [1–3]. China's offshore heavy oil resources are mainly distributed in the Bohai Bay, but the utilization degree of these heavy oil resources is less than 20%. This is because the conventional thermal recovery technology of heavy oil has poor adaptability on the sea, which is embodied in four aspects. Firstly, there is serious heat loss during steam injection, which makes the heat insufficient after steam injection into the formation. Secondly, the steam generation process requires a large amount of fresh water. Sea water desalination and oilfield production sewage treatment devices on offshore platforms are bulky, requiring a large amount of platform space and high treatment costs. Thirdly, the steam generation process of offshore platforms is complex; there are many steam generation and injection pipelines and there are prominent pipeline leakage and corrosion problems. Finally, the steam preparation process of offshore platforms is highly dependent on diesel, with high costs and a large amount of carbon dioxide emissions. The whole process is not low-carbon or environmentally friendly. In view of these problems, Zhou et al. [4] combined supercritical water technology with heavy oil thermal recovery technology, and creatively proposed the heavy oil recovery technology of supercritical multi-thermal fluid.

The main components of supercritical multi-thermal fluid are supercritical water, CO_2, and N_2. After the technology was proposed, some scholars carried out several studies on the generation of supercritical multi-thermal fluid. These studies suggested that the generation process of supercritical multi-thermal fluid could be divided into two steps. First, with the strong dissolution and diffusion properties of supercritical water, various organic waste liquids are fully dissolved. Then, they are burned with oxygen-containing gas dissolved in supercritical water to produce supercritical multi-thermal fluid [4–6].

As supercritical multi-thermal fluid is produced by the reaction of supercritical water with organic matters such as oil, the research on the reaction of supercritical water with organic matters is also of reference value. These studies focus on the oxidation process of organic matters by supercritical water. According to research results, supercritical water has the effects of solvent and dispersion, hydrogen transfer, and acid catalysis, which can improve the conversion rate of raw materials, reduce the coking amount of the system, and has obvious effects of desulfurization, denitrification, and heavy metal removal [7,8]. In the process of supercritical water oxidation, the organic waste liquid is completely degraded by a rapid oxidation reaction with oxidants such as O_2 in the supercritical water medium, which can realize self-heating and reduce energy consumption [9–12].

When the concentration of organic matter reaches a certain value, the heat released by the reaction can be used to maintain the heat balance of the process, so as to realize the energy self-compensation reaction and improve the economy of the whole process. Experimental research found that when the oxygen margin was greater than 10%, any substance could be completely separated, and the increase in the amount of oxygen had no effect on the reaction [13]. When the concentration of organic matter in the supercritical water oxidation reaction was greater than 2%, the autothermal reaction could be realized due to the release of heat during the reaction process [14–18]. These studies mainly focused on the supercritical water treatment of wastewater, and research on the supercritical water reaction mechanism and products after the reaction are relatively mature.

Some researchers have shown high interest in the application of supercritical water to heavy oil recovery, and carried out a series of studies. A heavy oil reservoir in northwest China took the lead in carrying out supercritical water huff and puff production tests [19]. On this basis, relevant research has been successively carried out. Through experimental research and theoretical analysis [20–22], the process, characteristics, and production of heavy oil recovery by supercritical water huff and puff technology have been discussed. Some studies focused on the process of supercritical water hydrothermal cracking of heavy oil, and discussed its reaction characteristics [23,24]. The research results showed that supercritical water, due to its high reaction characteristics, could split the macromolecular hydrocarbons in heavy oil into small molecules and produce a certain amount of coke. In this technology, research on the mechanism of enhancing heavy oil recovery by supercritical water had gradually emerged [25–28], and some of them [26] even started to conduct numerical simulations for this process. As supercritical water has already been used in the oil field to recover heavy oil, research on oilfield applications has gradually emerged. These studies mainly focus on the flow parameters and heat and mass transfer laws of supercritical water in the wellbore, and are discussed for vertical wells and horizontal wells, respectively [29–33]. Through research, the flow law and thermal diffusion law of supercritical water in the wellbore have been clarified, and the prediction model of physical parameters of supercritical water in the wellbore has been established.

In the last two years, researchers began to shift their attention to supercritical multi-thermal fluid. Based on the previous research on supercritical water, researchers gradually began to discuss the numerical simulation of supercritical multi-thermal fluid, the mechanism of enhanced oil recovery, and the cracking of heavy oil [34–36]. In general, the existing research provides a certain amount of basic knowledge for the research of supercritical multi-thermal fluid technology, and provides references for this research.

However, the existing research has hardly discussed the influence of different reactants and different temperature and pressure conditions on supercritical multi-thermal fluid and its generation process, and has not considered it with the displacement effect. These are the questions that engineers care about in oilfield applications. This paper aimed to address these problems to better apply the supercritical multi-thermal fluid technology to oilfield applications, using generation and flooding experiments of supercritical multi-thermal fluid. The generation and feasibility of supercritical multi-thermal fluid is discussed by analyzing and comparing the generation process, the composition of products, enthalpy value, oil displacement efficiency, and other indicators of supercritical multi-thermal fluid under different conditions. The technical roadmap and organization of this study are shown in Figure 1. The research results of this paper have important reference value and significance in guiding the field applications of supercritical multi-thermal fluid.

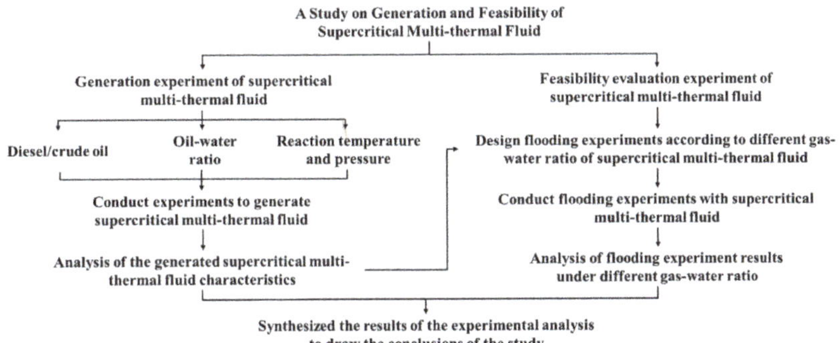

Figure 1. General sketch of this study.

2. Materials and Methods

The experiments in this study include supercritical multi-thermal fluid generation and flooding experiments. The experimental workflow is shown in Figure 2. The following will be introduced in accordance with the experimental workflow.

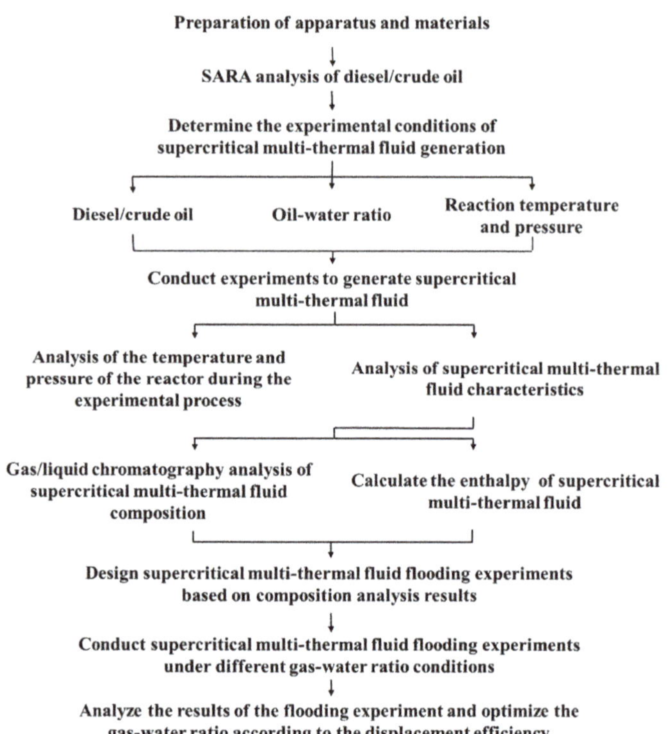

Figure 2. Workflow of experiments in this study.

2.1. Material

The diesel used in the experiment was 0# diesel produced by Sinopec (Figure A1), and the crude oil came from the L block in Bohai Bay of China (Figure A2). The viscosity of the crude oil at reservoir temperature (50 °C) was 4372.6 mPa, which belonged to ordinary heavy oil, and its viscosity–temperature relationship curve is shown in Figure 3. The porous medium in the one-dimensional flooding experiment was filled with 80-mesh quartz sand.

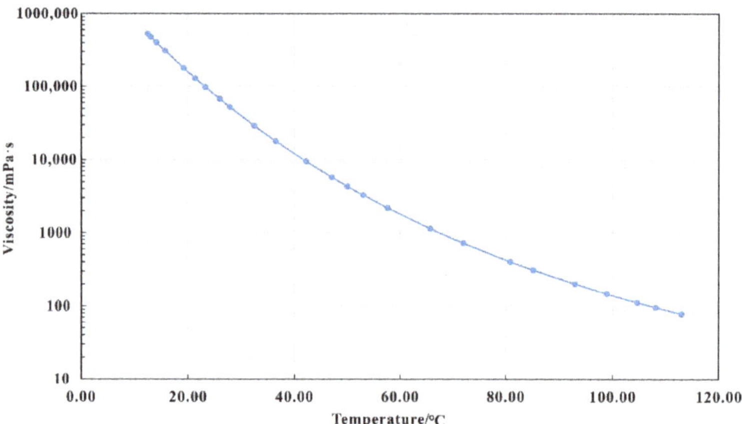

Figure 3. Viscosity–temperature relationship curve of crude oil from L block.

2.2. Apparatus and Procedures

2.2.1. Generation Experiment of Supercritical Multi-Thermal Fluid

(1) Apparatus

Supercritical multi-thermal fluid generation experiments were used to analyze the supercritical multi-thermal fluid generated by different proportions of an oil-water mixture under different conditions and its characteristics. As shown in Figure 4, the experimental apparatus mainly consists of a supercritical multi-thermal fluid generation system and chromatographic analysis system. The supercritical multi-thermal fluid generation system consisted of a high temperature and high pressure reactor (Figure A1), an ISCO pump, a unidirectional valve, and an intermediate container; the chromatographic analysis system consisted of a gas chromatograph and liquid chromatograph.

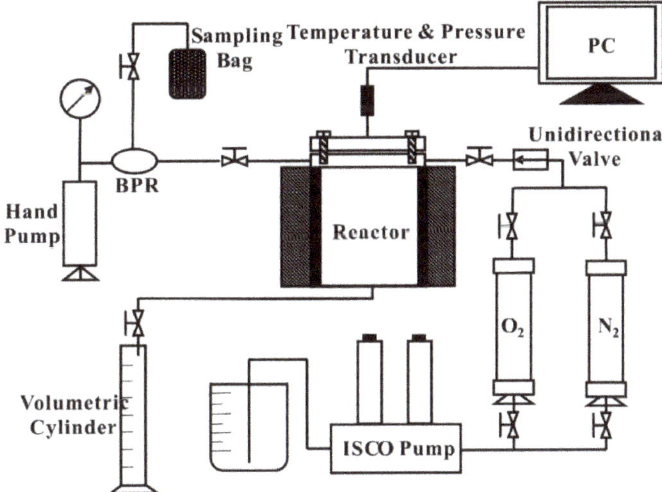

Figure 4. The flow chart of supercritical multi-thermal fluid generation experiment.

(2) Procedures

① Analyzed diesel/crude oil by liquid chromatography;

② Checked whether the experimental apparatus was working normally and whether the temperature and pressure sensors were in good condition, and repaired and replaced them in time if there were any problems;

③ Connected each experimental apparatus according to the experimental flow chart;

④ The diesel/crude oil and water were placed in the high temperature and high pressure reactor according to a predetermined ratio, and the reactor was strictly sealed. The high-temperature and high-pressure reactor used in the experiment was made of alloy HC276. The maximum working pressure was 40 MPa and maximum working temperature was 600 °C.

⑤ Opened the inlet and outlet valves, charged nitrogen gas through the inlet with a large flow for more than 10 min to ensure that the air in the high-temperature and high pressure reactor was completely exhausted, and then closed the outlet valve;

⑥ Nitrogen was filled into the high-temperature and high-pressure reactor at the predetermined pressure, then the power supply of the reactor was turned on, and the heating mode was chosen;

⑦ After the temperature and pressure in the reactor reached supercritical condition, the heating was stopped, and the shaking was started. At the same time, oxygen was injected into the reactor until the reaction ended, and the temperature and pressure in the reactor were monitored and recorded in real time during the whole reaction process;

⑧ After the reaction was completed, the reactor was left to cool down, and after the temperature in the reactor dropped to room temperature, the outlet was opened to collect gas samples in the reactor. Then the pressure was released and the reactor was opened to collect the liquid and solids (if any) in the reactor;

⑨ The collected gas–liquid samples were analyzed by the chromatograph.

2.2.2. Evaluation Experiment of Supercritical Multi-Thermal Fluids with Different Compositions

(1) Apparatus

The one-dimensional flooding experiment of supercritical multi-thermal fluid was used to evaluate and compare the displacement efficiency of supercritical multi-thermal fluid with different compositions in a single sand pack. As shown in Figure 5, the experimental apparatus was mainly composed of three parts: an injection and production system, model system, and monitoring system. Among them, the injection and production system was composed of an ISCO pump, supercritical water generator, gas mass flow controller, back pressure valve, hand pump, and a group of band heaters; the model system included an air bath, sand pack, and several buffer containers; the monitoring system included a computer and five groups of temperature and pressure sensors.

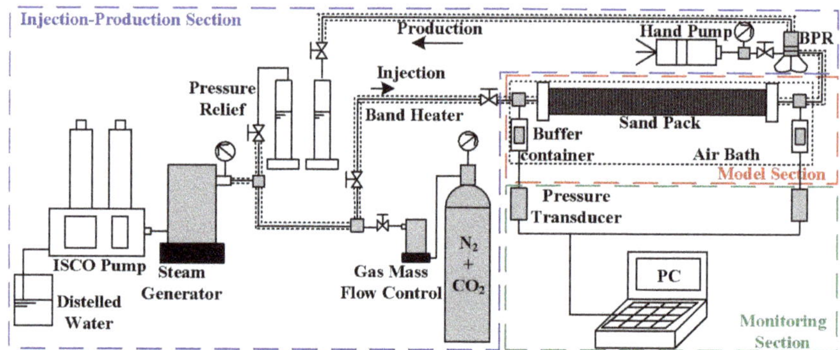

Figure 5. Flow chart of one-dimensional flooding experiment.

(2) Procedures

The one-dimensional flooding experiment mainly included the following steps:

① The experiment was carried out in accordance with the industry standard SY/T 6315-2017 "Determination Method for High-Temperature Relative Permeability and Oil Displacement Efficiency of Heavy Oil Reservoirs";

② The sand pack was filled with 80-mesh quartz sand, saturated with water after vacuuming, and porosity of the sand pack was calculated according to the volume of water inflow;

③ The experimental apparatus was connected according to the experimental flow chart, flooding the sand pack with different flow rates of water, recording the pressure difference between the two ends of the sand pack, and then calculating the permeability of the sand pack according to Darcy's law;

④ Saturating the sand pack with crude oil at a low flow rate, irreducible water saturation was calculated according to the volume of outflowing water, and the sand pack was placed in an air bath to age for more than 24 h;

⑤ After completing model initialization, the experimental running phase began. We injected supercritical steam or multi-thermal fluid into the sand pack according to the predetermined flow, and controlled the back pressure of the outlet to maintain it above 22.5 MPa;

⑥ During the experiment, the produced liquid was collected in real time, and the production time and produced oil and water volume were recorded, which was convenient for calculating the injected PV and degree of recovery. When the water cut was greater than 98%, the experiment was stopped to calculate and determine the oil displacement efficiency of supercritical multi-thermal fluid with different compositions. On this basis, the reasonable composition of supercritical multi-thermal fluid was determined;

⑦ After the experiment, the sand pack was opened and the distribution of remaining oil was observed.

3. Results

3.1. Generation Experiments of Supercritical Multi-Thermal Fluid

3.1.1. Supercritical Water–Diesel Reaction Process under Different Conditions

Ten supercritical water–diesel reaction experiments were carried out under different oil–water ratios or different temperature and pressure conditions. Based on the experimental process, reaction characteristics were analyzed. The experimental parameters are shown in Tables 1 and 2.

Table 1. Experimental parameters of supercritical water–diesel reactions to generate supercritical multi-thermal fluid under different temperature and pressure conditions.

No.	Reactant	Initial Temperature and Pressure
1	45 mL water + 5 mL diesel oil	400 °C/23 MPa
2	45 mL water + 5 mL diesel oil	400 °C/25 MPa
3	45 mL water + 5 mL diesel oil	450 °C/25 MPa
4	45 mL water + 5 mL diesel oil	500 °C/25 MPa
5	45 mL water + 5 mL diesel oil	500 °C/23 MPa

Table 2. Experimental parameters of supercritical water–diesel reactions to generate supercritical multi-thermal fluid under different oil–water ratios.

No.	Reactant	Initial Temperature and Pressure
1	45 mL water + 5 mL diesel oil (10% diesel oil)	400 °C/25 MPa
2	28.3 mL water + 5 mL diesel oil (15% diesel oil)	400 °C/25 MPa
3	20 mL water + 5 mL diesel oil (20% diesel oil)	400 °C/25 MPa
4	15 mL water + 5 mL diesel oil (25% diesel oil)	400 °C/25 MPa
5	11.67 mL water + 5 mL diesel oil (30% diesel oil)	400 °C/25 MPa

According to the experimental steps, the SARA analysis of the 0# diesel oil used was carried out before the experiments. The analysis results are shown in Table 3. During the experiment, the temperature and pressure changes in the reactor were recorded. After preheating to the predetermined temperature and pressure, the heating was turned off, and the reaction was started. After the experiment, when the reactor had been cooled to room temperature, the gas in the reactor was collected for gas chromatographic analysis, and the remaining liquid was collected for separation and analysis.

Table 3. Results of SARA analysis of 0# diesel oil.

Component	Value
Asphaltenes	/
Resins	4.4692%
Aromatics	9.9341%
Saturates	85.5967%
Total	100%

According to the results of the chromatographic analysis of the gas collected after the reaction, under the condition of sufficient oxygen, supercritical water and diesel were completely reacted, and there was no residual organic matter; the remaining gases were carbon dioxide, nitrogen and, a small amount of oxygen, and the remaining liquid was pure water, as shown in the Figure 6.

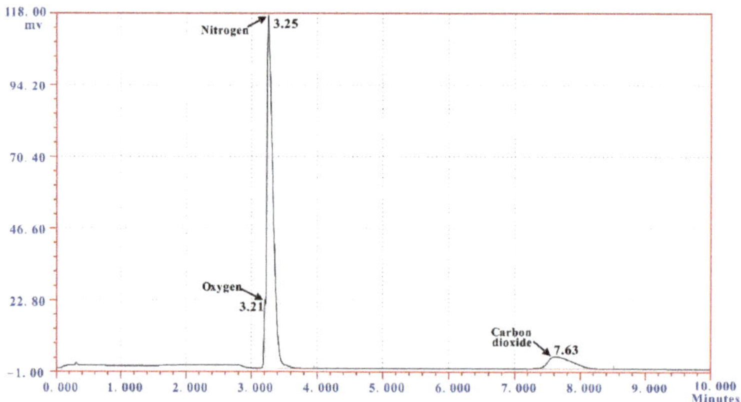

Figure 6. Analysis results of gas components collected after the reaction.

Judging from the temperature and pressure changes in the reactor during the supercritical water–diesel reaction experiment with different oil–water ratios, as shown in Figures 7 and 8, changes in the temperature and pressure under different conditions were basically the same, and could be roughly divided into three stages. In the first stage, after oxygen was injected into the reactor, the reactants rapidly reacted, and released a large amount of heat, so that the temperature in the reactor sharply rose, and the pressure in the reactor also obviously increased. In the second stage, with the progression of the reaction, the temperature in the reactor rapidly dropped after reaching the highest point, indicating that the reactants were decreasing at this time, heat release of the reaction was reduced, and heat released by the reactants per unit time was less than heat dissipation of the reactor per unit time. At this stage, the pressure in the reactor significantly fluctuated. In the third stage, the reactants were completely consumed. The reaction was completely finished, no heat was released, the temperature in the reactor dropped to slightly higher than the initial temperature, and the temperature tended to remain unchanged. The internal temperature tended to be stable, and the volume of the mixing system in the reactor tended to be stable as well, so that the pressure in the reactor was kept constant.

Combined with the results of chromatographic analysis, it was recognized that, under the condition of sufficient oxygen, diesel oil could completely react in each experiment to generate supercritical multi-thermal fluid. However, due to the different amounts of nitrogen that was injected to achieve the same initial reaction pressure under the conditions of different oil–water ratios, the gas composition of the supercritical multi-thermal fluid generated in each experiment was slightly different. This was mainly because the lower the water ratio, the more nitrogen was injected, and the more nitrogen present in the supercritical multi-thermal fluid generated by the reaction.

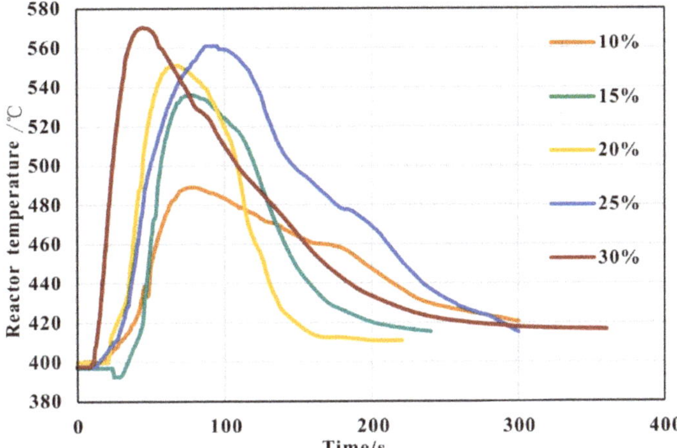

Figure 7. Temperature changes in the reactor during the supercritical water–diesel reaction experiment under different oil–water ratios.

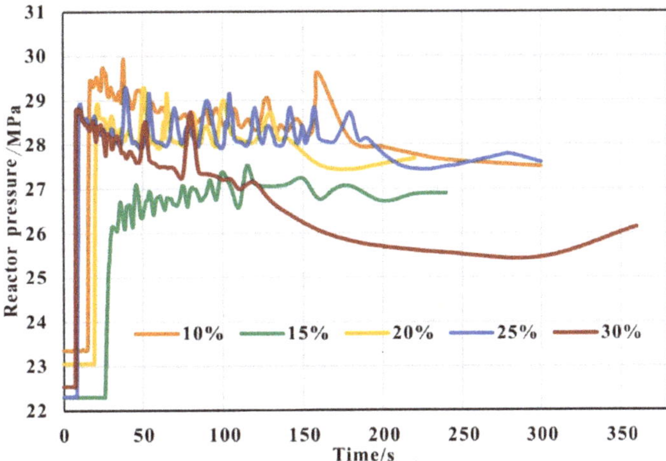

Figure 8. Pressure changes in the reactor during the supercritical water–diesel reaction experiment under different oil–water ratios.

When the experimental processes of different oil–water ratios were compared, it was apparent that the higher the diesel ratio, the higher the maximum temperature that could be achieved in the reactor. On the contrary, when the proportion of diesel oil was low, the amount of water was relatively higher. As the heat capacity of water is large and the amount of diesel oil was fixed, the heat released by the reaction was fixed as well, and the system needed more heat to raise a certain temperature. Thus, the maximum temperature in the reactor was lower. At the same time, the higher the proportion of supercritical water in the product, the greater the heat capacity, the less condensation during the cooling process, and the longer the pressure could be maintained.

Comparing the experimental process of different initial temperature and pressure conditions, Figures 9 and 10 show that the change laws of temperature and pressure were basically consistent with the experiments of different oil–water ratios, and could also be roughly divided into three stages. However, in experiments with different initial temperature and pressure conditions, the temperature rise in each experiment was very similar. Considering that the proportion of oil and water in each experiment, total amount of reactants, and the composition and total amount of the products were basically the same, it could be inferred that the reaction heat would be similar. Compared to experiments with different oil–water ratios, the pressure change curves in the reactor were very close in each experiment with different initial temperature and pressure conditions, and the final stable values were basically the same. Thus, the initial temperature and pressure conditions had little effect on the reaction.

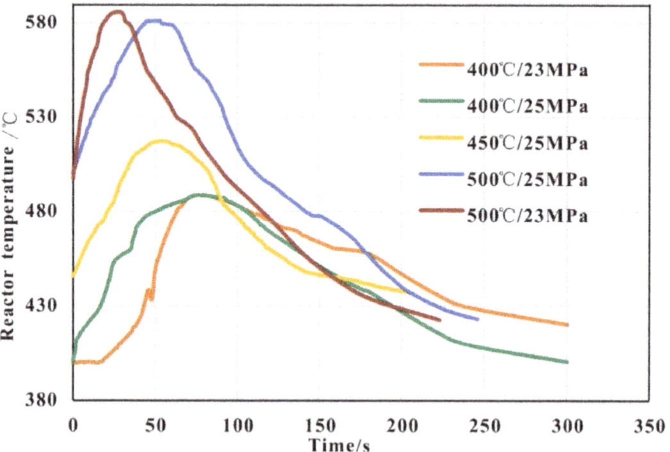

Figure 9. Temperature changes in the reactor during the experiment of supercritical water–diesel reactions under different initial temperature and pressure conditions.

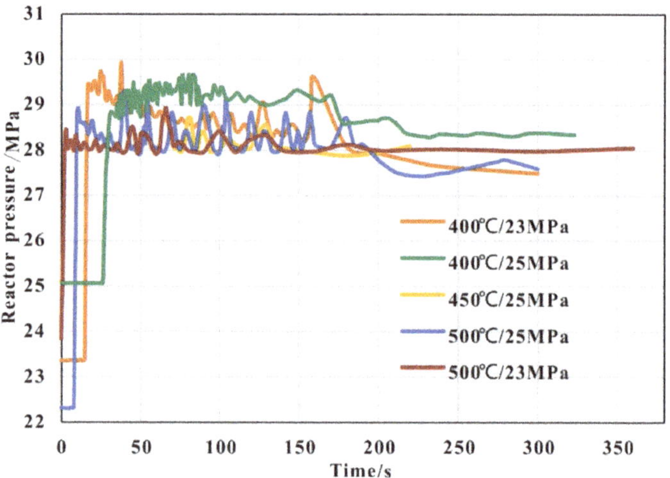

Figure 10. Pressure changes in the reactor during the experiment of supercritical water–diesel reactions under different initial temperature and pressure conditions.

3.1.2. Supercritical Water–Crude Oil Reaction Process under Different Conditions

Ten experiments of supercritical water–crude oil reactions with different oil–water ratios and different temperature and pressure conditions were conducted, and the reaction characteristics were analyzed on the basis of the experimental process. The experimental method and data analysis were the same as the previous supercritical water–diesel reaction experiments. The parameters determined for the experiment are shown in Tables 4 and 5.

Table 4. Experimental parameters of supercritical water–crude oil reaction to generate supercritical multi-thermal fluids under different temperature and pressure conditions.

No.	Reactant	Initial Temperature and Pressure
1	45 mL water + 5 mL crude oil	400 °C/23 MPa
2	45 mL water + 5 mL crude oil	400 °C/25 MPa
3	45 mL water + 5 mL crude oil	450 °C/25 MPa
4	45 mL water + 5 mL crude oil	500 °C/25 MPa
5	45 mL water + 5 mL crude oil	500 °C/23 MPa

Table 5. Experimental parameters of supercritical water–crude oil reaction to generate supercritical multi-thermal fluid under different oil–water ratios.

No.	Reactant	Initial Temperature and Pressure
1	45 mL water + 5 mL crude oil(10% crude oil)	400 °C/25 MPa
2	28.3 mL water + 5 mL crude oil(15% crude oil)	400 °C/25 MPa
3	20 mL water + 5 mL crude oil(20% crude oil)	400 °C/25 MPa
4	15 mL water + 5 mL crude oil(25% crude oil)	400 °C/25 MPa
5	11.67 mL water + 5 mL crude oil(30% crude oil)	400 °C/25 MPa

SARA analysis of crude oil was conducted before the experiment (Table 6), and the experimental apparatus and procedures were the same as above.

Table 6. Results of SARA analysis of crude oil.

Component	Value
Asphaltenes	25.9861%
Resins	8.1372%
Aromatics	19.6532%
Saturates	46.2236%
Total	100%

Similar to the supercritical water–diesel reaction experiments, under the condition of sufficient oxygen, the supercritical water–crude oil completely reacted, with no organic matter remaining, remaining gases including carbon dioxide, nitrogen, a small amount of oxygen, a small amount of sulfur dioxide, and other gases (Figure 11), and the remaining liquid being pure water (Figure A6). In other words, the reaction of supercritical water and crude oil could also generate supercritical multi-thermal fluid

By comparing the experimental process of different oil–water ratios, the water–crude oil reaction process was found to be similar to the water–diesel oil reaction process, and the temperature and pressure changes in the reactor (Figures 12 and 13) could also be divided into three stages: ① The temperature and pressure rapidly rose; ② The temperature sharply dropped after reaching the maximum value, and the pressure obviously fluctuated; ③ The temperature gradually decreased and tended to be stable; the pressure tended to be stable. At the same time, the higher the proportion of crude oil in the crude oil reaction process, the higher the maximum temperature that could be achieved in the reactor. On the contrary, the calorific value of crude oil was lower than that of diesel oil, the temperature rise in the reactor was lower than that of diesel oil, and there was less heat release in the water–crude oil reaction.

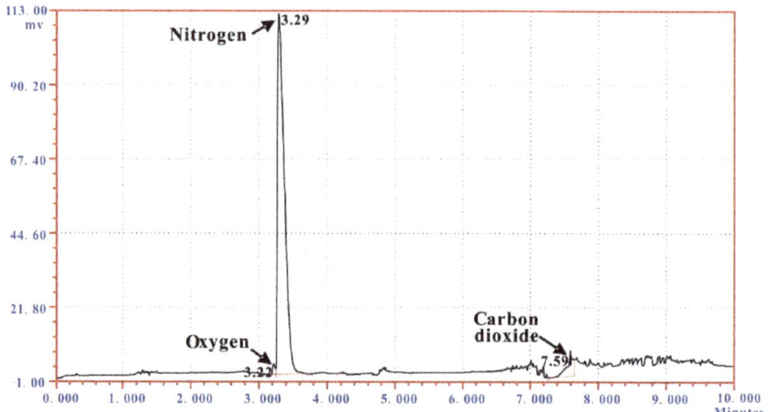

Figure 11. Analysis results of gas components collected after the reaction.

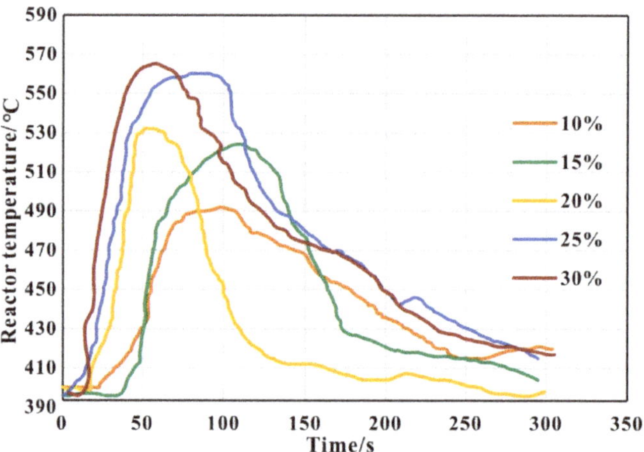

Figure 12. Temperature changes in the reactor during the supercritical water–crude oil reaction experiment under different ratios of oil and water.

The experimental process of different initial temperature and pressure conditions indicated that the water–crude oil reaction process was similar to that of the water–diesel reaction, and temperature increases in the reactor were also very close, as shown in Figures 14 and 15. Considering that the proportion of oil and water in each experiment was kept constant, the total amount of reactants was the same, and the composition and total amount of the products were basically the same, it could also be inferred that the reaction heat would be similar, and that initial temperature and pressure conditions had relatively little influence on the reaction.

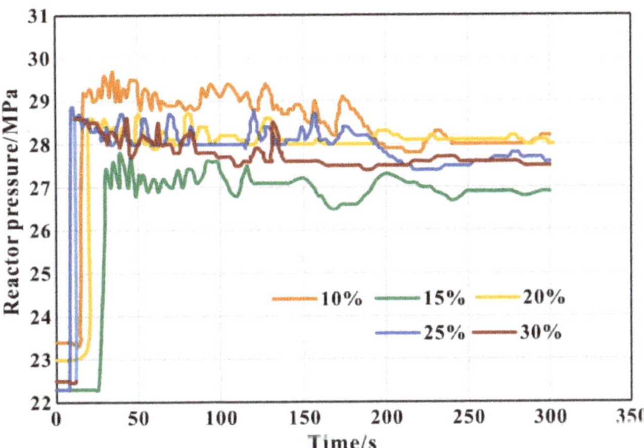

Figure 13. Changes of pressure in the reactor during the supercritical water–crude oil reaction experiment under different oil–water ratios.

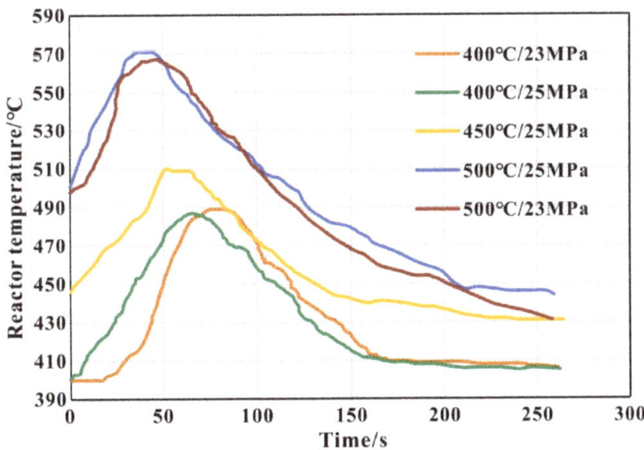

Figure 14. Temperature changes in the reactor during the experiment of supercritical water–crude oil reaction under different initial temperature and pressure conditions.

3.1.3. Comparison of Supercritical Multi-Thermal Fluid Generated under Different Conditions

Component Composition

As mentioned above, after carrying out supercritical water–oil (diesel oil or crude oil) reaction experiments under different oil–water ratios and different temperature and pressure conditions, the reactor was cooled to room temperature. We then collected gas samples in the reactor and conducted chromatographic analysis, to determine the composition of the supercritical multi-thermal fluid generated by the reaction.

The chromatographic analysis results (Tables 7 and 8) showed that the main components of gas products (at room temperature) after the reaction of diesel oil and heavy oil were similar, that is, they mainly comprised carbon dioxide, nitrogen, and oxygen not consumed by the reaction. However, compared with the reaction products of diesel oil, the reaction products of heavy oil contained more impure gases, such as sulfur dioxide and nitrogen dioxide. This was because, compared with diesel oil, crude oil contains asphaltene, which contains non-hydrocarbon substances with heteroatoms such as sulfur and nitrogen, converting into sulfur dioxide and nitrogen dioxide after the full reaction under conditions of sufficient oxygen.

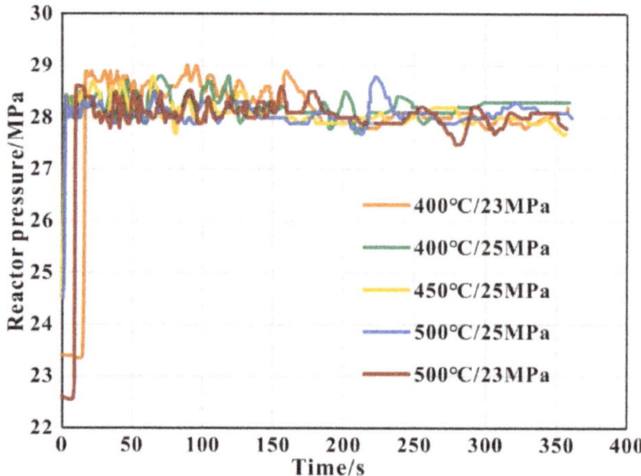

Figure 15. Changes of pressure in the reactor during the experiment of supercritical water–crude oil reaction under different initial temperature and pressure conditions.

Table 7. Gas chromatographic analysis results after supercritical water–diesel reaction.

No.	Type	Nitrogen,%	Carbon Dioxide, %	Oxygen, %
1	10% diesel	66.52	11.52	21.96%
2	15% diesel	66.70	10.43	22.87%
3	20% diesel	63.53	10.47	26.00%
4	25% diesel	64.90	8.84	26.26%
5	30% diesel	67.33	8.59	24.08%
6	400 °C/ 23 MPa	66.52	11.52	21.96
7	400 °C/ 25 MPa	65.31	10.48	24.21
8	450 °C/ 25 MPa	65.77	11.03	23.20
9	500 °C/ 25 MPa	69.42	12.07	18.51
10	500 °C/ 23 MPa	66.70	13.01	20.29

Table 8. Gas chromatographic analysis results after supercritical water–crude oil reaction.

No.	Type	Nitrogen,%	Carbon Dioxide, %	Oxygen, %	Sulfur Dioxide, %	Nitrogen Dioxide, %
1	10% crude oil	63.79	12.40	23.43	0.12	0.27
2	15% crude oil	61.68	11.03	26.81	0.15	0.33
3	20% crude oil	66.57	11.50	21.66	0.08	0.19
4	25% crude oil	63.06	8.04	28.32	0.18	0.40
5	30% crude oil	64.19	8.43	26.86	0.16	0.36
6	400 °C/ 23 MPa	63.79	12.40	23.43	0.12	0.27
7	400 °C/ 25 MPa	66.21	11.74	21.81	0.07	0.17
8	450 °C/ 25 MPa	63.10	11.95	24.37	0.18	0.40
9	500 °C/ 25 MPa	64.19	12.85	22.47	0.15	0.34
10	500 °C/ 23 MPa	62.60	13.84	22.76	0.24	0.55

When the reaction products under different initial temperature and pressure conditions were compared, the composition of reaction products were basically unchanged within the temperature and pressure range given in the experiment, whether the reactant was heavy oil or diesel oil. In other words, within the temperature and pressure range of these experiments, the influence of temperature and pressure on the reaction products could be ignored.

In contrast, the proportion of oil and water in the reactant had a more obvious effect on chromatography results. The total amount of oil was controlled to remain unchanged in the experiment, and the oil–water ratio was adjusted by changing the amount of water. The volume of the reactor could not be changed. In order to reach the same initial pressure, we varied the amount of nitrogen charged. The higher the proportion of oil, the less the amount of water, and the smaller the total volume of oil–water; meanwhile, to reach the same initial pressure as the other oil–water ratio experiments, the more nitrogen was charged. Therefore, the chromatogram of the gas collected after the reaction indicated that, with an increasing proportion of oil, the proportion of carbon dioxide decreased and the proportion of nitrogen increased. Due to the amount of oil involved in the reaction remaining constant, the amount of oxygen charged was roughly the same; thus, the proportion of oxygen in the chromatographic results was basically the same.

21) Enthalpy of Supercritical Multi-Thermal Fluids

Generally speaking, the higher the enthalpy of thermal fluid, the higher the heat it carried, which was more conducive to the recovery of heavy oil. Therefore, the enthalpy of the thermal fluid generated by the reaction was a very noteworthy index. It should be pointed out that, due to the experimental methods and processes used in this study, the proportion of nitrogen and oxygen in the reactor of the experiment was not the same as the proportion of nitrogen and oxygen in the air. To make the enthalpy value of supercritical multi-thermal fluid determined by calculation have universal significance and comparability, we corrected the amount of nitrogen in the reactor based on the chromatographic analysis results of the produced fluid, combined with the nitrogen–oxygen ratio in the air. At the same time, without considering the remaining oxygen in the reaction, the enthalpy h of the supercritical multi-thermal fluid produced in different reactions could be calculated. The calculation formula was as follows:

$$h = \frac{w_i}{\sum\limits_{i=1}^{n} w_i} h_i$$

Based on the specific enthalpy of thermal fluids, as shown in Tables 9 and 10, the lower the proportion of diesel oil and the higher the proportion of water, the higher the specific enthalpy of thermal fluids, because the thermal fluids were mainly carried by supercritical water, and the higher the proportion of supercritical water, the higher the enthalpy of supercritical multi-thermal fluid. According to the previous method, the enthalpy of supercritical multi-thermal fluid generated by the reaction in experiments under different initial temperature and pressure conditions was calculated. As mentioned above, the initial temperature and pressure basically had no effect on the composition of reaction products, so the total enthalpy and specific enthalpy of thermal fluid generated under different initial temperature and pressure conditions were very similar after conversion to the same temperature and pressure (400 °C, 25 MPa), and the average values were 180,819.937 J and 1507.402 J/g, respectively.

As with the previous method, the enthalpy of supercritical multi-thermal fluid produced by the reaction of supercritical water–crude oil was calculated on the basis of the composition of the produced fluid, the results are shown in Tables 11 and 12. Based on the specific enthalpy of thermal fluid, the law was also similar to that of the reaction of supercritical water–crude oil; thus, the higher the proportion of water, the higher the proportion of supercritical water in the product, and the higher the specific enthalpy of thermal fluid.

Table 9. Enthalpy of supercritical multi-thermal fluids generated by supercritical water–diesel reaction under different oil–water ratios.

No.	Type	Nitrogen Quantity, g	CO_2 Quantity, g	Total Water Quality, g	Total Mass of Fluid, g	Total Enthalpy of Thermal Fluids, J (400 °C, 25 MPa)	Specific Enthalpy of Thermal Fluids, J/g (400 °C, 25 MPa)
1	10% diesel	55.821	16.446	49.928	122.195	217,958.660	1783.70
2	15% diesel	53.435	15.743	33.227	102.405	163,223.870	1593.91
3	20% diesel	52.437	15.499	24.868	92.804	136,095.959	1466.49
4	25% diesel	55.981	16.493	19.931	92.405	124,608.843	1348.51
5	30% diesel	53.768	15.841	16.402	86.011	111,136.352	1292.12

Table 10. Enthalpy of supercritical multi-thermal fluids generated by supercritical water–diesel reaction under different initial temperature and pressure conditions.

No.	Type	Nitrogen Quantity, g	CO_2 Quantity, g	Total Water Quality, g	Total Mass of Fluid, g	Total Enthalpy of Thermal Fluids, J (400 °C, 25 MPa)	Specific Enthalpy of Thermal Fluids, J/g (400 °C, 25 MPa)
1	400 °C/ 23 MPa	54.749	16.130	49.914	120.793	181,454.782	1502.20
2	400 °C/ 25 MPa	55.126	16.241	49.919	121.286	181,830.813	1499.19
3	450 °C/ 25 MPa	54.219	15.974	49.908	120.101	180,928.838	1506.47
4	500 °C/ 25 MPa	53.737	15.832	49.901	119.470	180,446.440	1510.39
5	500 °C/ 23 MPa	52.726	15.534	49.888	118.148	179,438.812	1518.76

Table 11. Enthalpy of supercritical multi-thermal fluid generated by supercritical water-crude oil reaction under different oil-water ratios.

No.	Type	Nitrogen Quantity, g	CO_2 Quantity, g	SO_2 Quantity, g	NO_2 Quantity, g	Total Water Quality, g	Total Mass of Fluids, g	Total enthalpy of Thermal Fluids, J (400 °C, 25 MPa)	Specific Enthalpy of Thermal Fluids, J/g (400 °C, 25 MPa)
1	10% crude oil	42.856	12.673	0.312	0.512	48.888	105.241	167,995.845	1596.30
2	15% crude oil	43.188	12.879	0.248	0.408	32.282	89.005	125,462.946	1409.62
3	20% crude oil	43.777	13.244	0.137	0.224	24.064	81.446	104,784.173	1286.55
4	25% crude oil	42.410	12.396	0.397	0.652	18.803	74.658	90,029.995	1205.90
5	30% crude oil	42.695	12.573	0.342	0.562	15.358	71.530	81,393.320	1137.89

Table 12. Enthalpy of supercritical multi-thermal fluid generated by supercritical water–crude oil reaction under different initial temperature and pressure conditions.

No.	Type	Nitrogen Quantity, g	CO_2 Quantity, g	SO_2 Quantity, g	NO_2 Quantity, g	Total Water Quality, g	Total Mass of Fluid, g	Total Enthalpy of Thermal Fluids, J (400 °C, 25 MPa)	Specific Enthalpy of Thermal Fluids, J/g (400 °C, 25 MPa)
1	400 °C 23 MPa	43.544	13.100	0.181	0.297	49.019	106.141	168,931.956	1591.58
2	400 °C 25 MPa	43.868	13.301	0.119	0.196	49.081	106.565	169,373.536	1589.39
3	450 °C 25 MPa	43.050	12.793	0.275	0.452	48.925	105.495	168,260.094	1594.96
4	500 °C 25 MPa	43.338	12.972	0.220	0.361	48.980	105.871	168,651.679	1592.99
5	500 °C 23 MPa	42.810	12.644	0.321	0.527	48.879	105.181	167,932.908	1596.61

According to the above method, the enthalpy of supercritical multi-thermal fluid generated by the reaction was calculated in experiments under different initial temperature and pressure conditions. Similarly, when converted to the same temperature and pressure (400 °C, 25 MPa), the total enthalpy and specific enthalpy of thermal fluid generated under different initial temperature and pressure conditions were very close, with average values of 168,630.035 J and 1593.106 J/g, respectively.

The above four tables indicate that the total enthalpy of supercritical multi-thermal fluid generated in the reaction experiment of diesel and supercritical water was higher than that of supercritical multi-thermal fluid generated in the reaction of heavy oil, but the relationship between the specific enthalpy of the two was just the opposite. Taking the reaction experiment of 10% diesel oil and 10% heavy oil as an example, both reactions started at 400 °C and 25 MPa. It was evident from the experimental results that although the volume of oil was 5 mL, more carbon dioxide and water were generated after the reaction of diesel, and the difference in the carbon dioxide generated was greater than the difference in the water generated. According to further calculations, as shown in Tables 13 and 14, the density of diesel oil was lower, and the mass of 5 mL diesel oil was slightly less than 5 mL heavy oil, but more than 95% of diesel oil was made up of hydrocarbons, and the content of saturated hydrocarbons accounted for 90% of all hydrocarbons, whereas the hydrocarbon content of crude oil used in the experiment was less than 70%, and the content of saturated hydrocarbons was about half of the saturated hydrocarbons in diesel oil. Compared with non-hydrocarbon substances, hydrocarbon substances consumed more oxygen, which made diesel oil consume more oxygen in the reaction process, and correspondingly generated more carbon dioxide and water. As oxygen came from the air, the amount of nitrogen in the thermal fluids would also increase, so that the gas mass, total mass, and total enthalpy of the thermal fluids generated by the 5 mL diesel reaction were greater. On the contrary, crude oil had low oxygen consumption, and correspondingly less nitrogen in thermal fluids and less gas mass in the thermal fluid generated by the reaction of 5 mL of crude oil. Thus, crude oil reactions had a lower total mass of thermal fluids, and lower total enthalpy. Due to the fact that thermal fluid mainly relies on supercritical water to carry heat, and the heat capacity of the gas is far lower than that of water, the higher proportion of water in the supercritical multi-thermal fluid generated by the crude oil reaction increased the specific enthalpy of the supercritical multi-thermal fluid generated by the crude oil reaction than that of the supercritical multi-thermal fluid generated by the diesel oil reaction. This characteristic not only appeared in the reaction experiment of 10% diesel oil and 10% heavy oil, but in the other experiments.

Table 13. Gas–water mass ratio of supercritical multi-thermal fluid generated by supercritical water–diesel reaction under different conditions.

No.	Type	Total Gas Mass, g	Water Mass, g	Gas–Water Ratio
1	10% diesel	72.267	49.928	1.3
2	15% diesel	69.178	33.227	1.8
3	20% diesel	67.936	24.868	2.4
4	25% diesel	72.474	19.931	3.2
5	30% diesel	69.609	16.402	3.7
6	400 °C/ 23 MPa	71.4	49.919	1.4
7	400 °C/ 25 MPa	71.3	49.919	1.4
8	450 °C/ 25 MPa	67.9	49.884	1.4
9	500 °C/ 25 MPa	69.1	49.897	1.4
10	500 °C/ 23 MPa	71.4	49.922	1.4

Table 14. Gas–water mass ratio of supercritical multi-thermal fluid generated by supercritical water crude oil reaction under different conditions.

No.	Type	Total Gas Mass, g	Water Mass, g	Gas–Water Ratio
1	10% crude oil	56.353	48.888	1.2
2	15% crude oil	56.723	32.282	1.8
3	20% crude oil	57.382	24.064	2.4
4	25% crude oil	55.855	18.803	3.0
5	30% crude oil	56.172	15.358	3.7
6	400 °C/ 23 MPa	57.122	49.019	1.2
7	400 °C/ 25 MPa	57.484	49.081	1.2
8	450 °C/ 25 MPa	56.570	48.925	1.2
9	500 °C/ 25 MPa	56.891	48.980	1.2
10	500 °C/ 23 MPa	56.302	48.879	1.2

3.2. *Evaluation Experiment of Supercritical Multi-Thermal Fluid Flooding*

The previous analysis of the supercritical multi-thermal fluid generation experiment showed that the gas–water mass ratio in the supercritical multi-thermal fluid generated by the reaction of different oil–water ratios was different. Existing field practice has confirmed that the more gas there is, the better the process of heavy oil. Therefore, it is necessary to further optimize the proportion of non-condensable gas and supercritical water in the thermal fluid injected underground, and determine the appropriate gas–water ratio that can better use supercritical multi-thermal fluid for heavy oil recovery.

Combined with the composition of supercritical multi-thermal fluid generated in the previous experiment and the mass ratio of gas–water, seven experiments with the mass ratio of gas–water in a range of 1–4 were carried out. The experimental parameters are shown in Table 15.

Table 15. One-dimensional experimental parameters.

No.	Porosity (%)	Permeability (D)	Oil Saturation (%)	Gas–Water Mass Ratio
1	36.8	3.95	86.2	1
2	38.4	4.59	84.3	1.5
3	37.3	4.24	85.1	2
4	36.7	4.45	87.2	2.5
5	36.9	4.29	83.7	3
6	35.4	3.91	84.2	3.5
7	36.3	4.19	86.3	4

During the experiment, with an increase in the gas–water mass ratio, the supercritical area was found to be narrowed (Figure 16), and the existence of non-condensable gas made gas channeling appear in the experiment, which then led to steam channeling. The higher the gas–water mass ratio, the earlier and more serious this phenomenon appeared. At the same time, the emulsification of the produced liquid was serious when the supercritical multi-thermal fluid flooding with a low gas–water mass ratio was used. With the increase of the gas–water mass ratio, the liquid emulsification of the products decreased. After opening the model and observing the oil sand, it could be seen that the further the quartz sand was from the injection end after flooding, the smaller the sweep multiple and the darker the color (Figure A7).

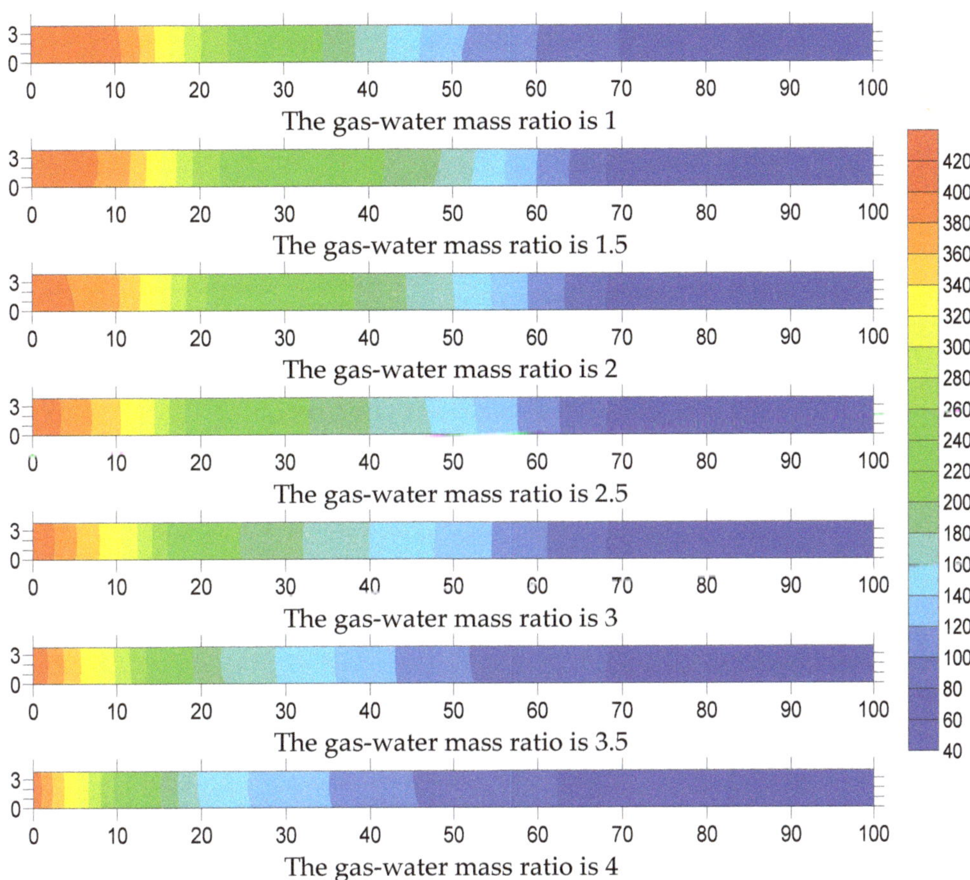

Figure 16. Real-time temperature distribution of supercritical multi-thermal fluid flooding.

Figures 17 and 18 demonstrate that in the early stage of flooding, the supercritical multi-thermal fluid with different gas–water mass ratios could achieve higher oil recovery rates, and with the progress of flooding, the oil recovery rate gradually slowed down. Moreover, the lower the gas–water mass ratio, the higher the oil recovery rate in the early stage, and overall oil recovery rate was also higher than that of the supercritical multi-thermal fluid with a high gas–water mass ratio. With an increase in non-condensable gas in the supercritical multi-thermal fluid, the oil displacement efficiency decreased. When the water content reached 98%, the oil displacement efficiency was 90.72, 88.8, 86.3, 84.2, 82.71, 80.30, and 71.8%. Therefore, from the perspective of oil displacement efficiency, in these five experiments, the gas–water mass ratio in the supercritical multi-thermal fluid was optimal when it was about 1. In other words, when the proportion of oil was about 10%, the supercritical multi-thermal fluid produced by the reaction had the best oil displacement effect.

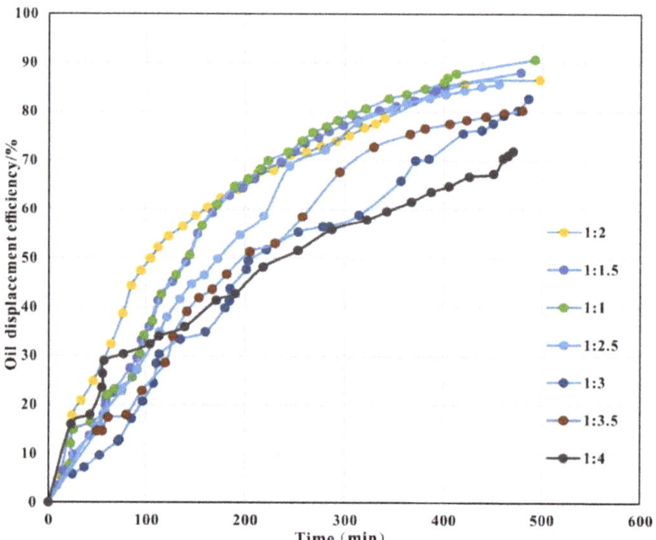

Figure 17. Oil displacement efficiency versus time.

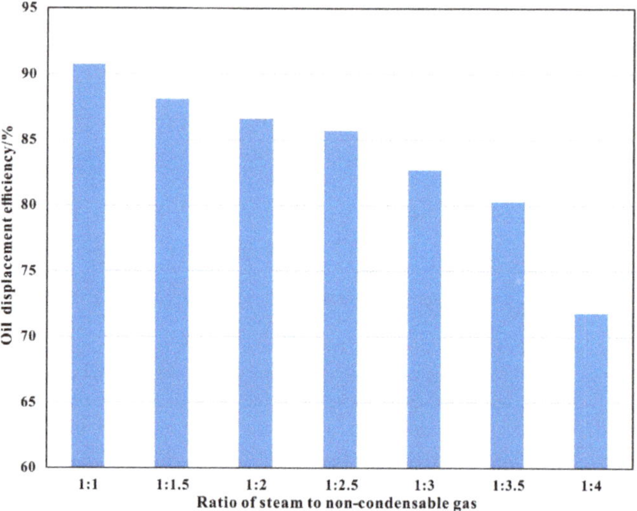

Figure 18. Oil displacement efficiency under different steam/non-condensable gas ratios.

4. Discussion

In this study, the generation and application of supercritical multi-thermal fluid were considered as a whole, and then the optimization direction and factors to be considered for supercritical multi-thermal fluid were clarified. Through this study, we clarified the influence of reactants and temperature and pressure conditions during the reaction on supercritical multi-thermal fluid and its generation process. We also clarified the sweep efficiency of supercritical multi-thermal fluid generated under different conditions, and the optimization of supercritical multi-thermal fluid was expounded.

Our study demonstrated that the proportion of oil and water in the generation process of supercritical multi-thermal fluid not only had a direct and obvious impact on the generated thermal fluid, but also indirectly affected the utility of the thermal fluid. In oilfield applications, the above aspects should be considered for the optimization of supercritical multi-thermal fluid.

On the one hand, whether diesel oil or heavy oil was used to react to generate supercritical multi-thermal fluid, an increase in the proportion of oil led to an increase in the proportion of gas in the generated supercritical multi-thermal fluid, that is, the gas–water mass ratio increased. This increase in the proportion of oil brings about an increase in the cost of supercritical multi-thermal fluid, whereas the increase in the gas–water mass ratio has no positive impact on the effect of supercritical multi-thermal fluid. Both previous practical experience and the experimental results of this study confirmed that when the gas–water mass ratio was relatively low, the thermal fluid recovery effect was better than when the gas–water mass ratio was high.

On the other hand, a low proportion of oil does not always result in the most optimal supercritical multi-thermal fluid generation. In fact, the goal of the supercritical multi-thermal fluid generation process is to create a suitable initial reaction condition to promote the reaction of supercritical water and oil under conditions of sufficient oxygen to generate supercritical multi-thermal fluid. The reaction process energy is no longer provided from the outside world, and the heat of reaction requires not only maintaining the high enthalpy value of the supercritical multi-component thermal fluid, but also heating the subsequent incoming reactants to continue the reaction. Thus, the proportion of oil must not be reduced too much in pursuit of a low gas–water mass ratio. When the gas–water mass ratio of thermal fluid is too low, it will become a single supercritical water, which is no longer a supercritical multi-thermal fluid.

The universal oil–water ratio really depends on whether diesel oil, crude oil, or a mixture of the two is used. At the same time, it also depends on the thermal efficiency of the supercritical multi-thermal fluid generator itself. In this study, this issue was not discussed, which is a limitation of this study. This issue should be further explored in future studies. Another limitation of the current study was that some of the experimental research and analysis appeared relatively rough, due to limits in experimental conditions, which is a disadvantage of this study that should be addressed in future research.

5. Summary and Conclusions

In order to study the generation and feasibility of supercritical multi-thermal fluid, generation and flooding experiments of supercritical multi-thermal fluid were carried out, respectively. From the analysis and discussion of the experimental results, the following conclusions were reached:

(1) In the process of supercritical multi-thermal fluid generation, the higher the proportion of oil in the reactant, the higher the maximum temperature that could be reached in the reactor. When the proportion of oil and water in the reactant was certain, the temperature rise in the reactor was basically the same under different initial reaction temperature conditions. The optimal proportion of oil and water in the reactant should be determined according to the specific oil and supercritical multi-thermal fluid generator.

(2) Within the scope of this study, the effect of the reaction initiation temperature on the generation of supercritical multi-thermal fluid could be neglected. In contrast, the proportion of oil and water in the reactant had a significant impact on the supercritical multi-thermal fluid generated. The higher the proportion of oil, the more gas in the generated supercritical multi-thermal fluid, the higher the mass ratio of gas–water, and the lower the specific enthalpy of the thermal fluid. Compared with diesel oil, the gas–water mass ratio of the supercritical multi-thermal fluid produced by the reaction of crude oil was lower and the specific enthalpy was higher under the same oil–water ratio.

(3) By carrying out the supercritical multi-thermal fluid flooding experiment, it was confirmed that the supercritical multi-thermal fluid with a low gas–water mass ratio

had higher oil displacement efficiency, higher oil recovery rate in the early stage, a larger supercritical area formed in the oil layer, and later channeling. In this study, when the gas–water mass ratio in supercritical multi-thermal fluid was about 1, the oil displacement efficiency of supercritical multi-thermal fluid was the highest, reaching 90.72%, and the corresponding oil–water ratio of reactants during the generation of supercritical multi-thermal fluid was about 10%.

(4) The proportion of oil and water in the generation process of supercritical multi-thermal fluid directly affected the composition and specific enthalpy of the generated thermal fluid, which then affected the application effect of thermal fluid. In oilfield applications, the above factors should be taken into account for the optimization of the generation process of supercritical multi-thermal fluid.

Author Contributions: X.T., J.Z. and Q.F. conceived and designed the experiments; J.T. conducted the experiments; Z.H. and J.T. analyzed the data; J.T. wrote the paper. All authors have read and agreed to the published version of the manuscript.

Funding: This research was funded by the State Key Laboratory of Offshore Oil Exploitation in China [grant number CCL2021RCPS0508KQN].

Institutional Review Board Statement: Not applicable.

Informed Consent Statement: Not applicable.

Data Availability Statement: The data that support the findings of this study are available from the corresponding author.

Acknowledgments: The authors would like to acknowledge the open fund project from State Key Laboratory of Offshore Oil Exploitation in China for providing financial support and oil samples.

Conflicts of Interest: The authors declare no conflict of interest.

Nomenclature

h	Enthalpy of the supercritical multi-thermal fluid (J)
h_i	Enthalpy of component i in the supercritical multi-thermal fluid (J)
w_i	Mass of component i in the supercritical multi-thermal fluid (g)
I	Refers to any component in the supercritical multi-thermal fluid

Appendix A

Figure A1. 0# diesel.

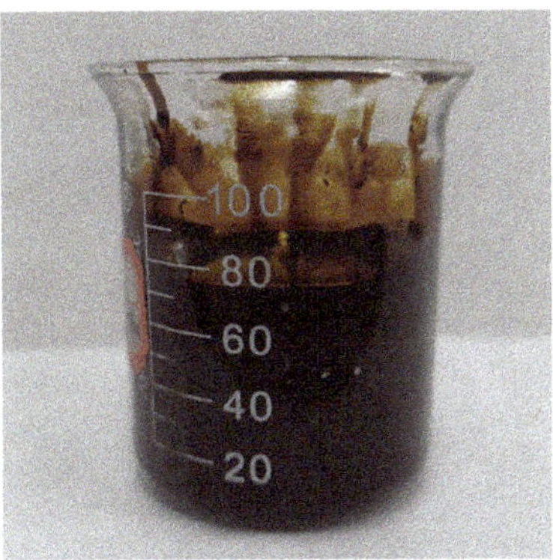

Figure A2. Crude oil of L block.

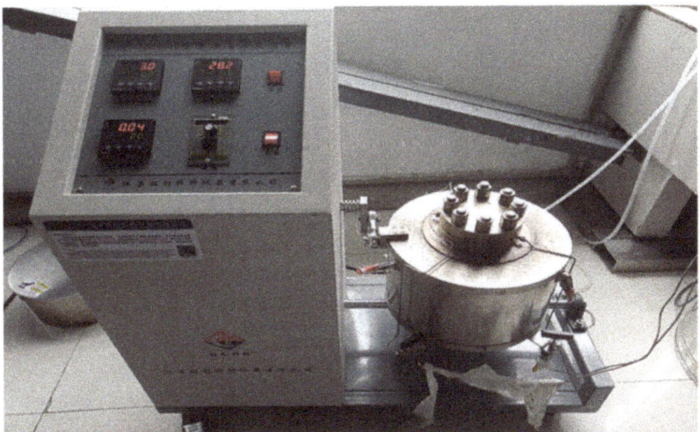

Figure A3. High temperature and high pressure reactor for the generation experiments.

Figure A4. One-dimensional sand pack.

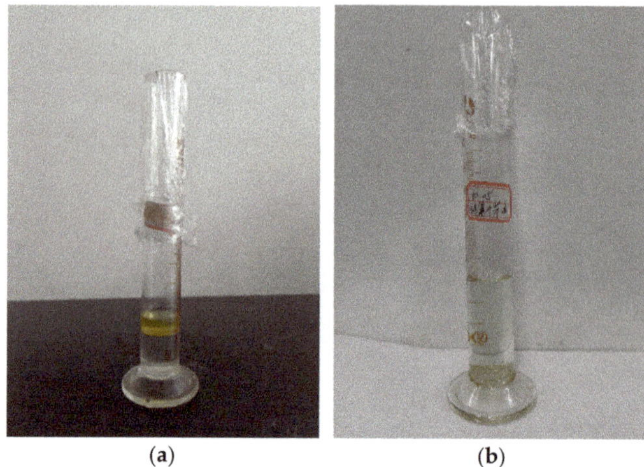

Figure A5. Diesel oil–water mixture (**a**) before reaction; (**b**) after reaction.

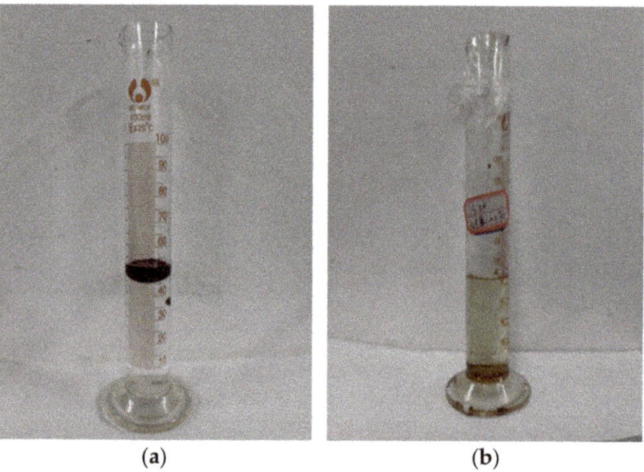

Figure A6. Crude oil–water mixture (**a**) before reaction; (**b**) after reaction.

Figure A7. Oil sands after one-dimensional flooding.

References

1. Dong, X.; Liu, H.; Chen, Z.; Wu, K.; Lu, N.; Zhang, Q. Enhanced oil recovery techniques for heavy oil and oil sands reservoirs after steam injection. *Appl. Energy* **2019**, *239*, 1190–1211. [CrossRef]
2. Lu, C.; Liu, H.; Zhao, W.; Lu, K.; Liu, Y.; Tian, J.; Tan, X. Experimental Investigation of In-Situ Emulsion Formation to Improve Viscous-Oil Recovery in Steam-Injection Process Assisted by Viscosity Reducer. *SPE J.* **2017**, *22*, 130–137. [CrossRef]
3. Lyu, X.; Liu, H.; Pang, Z.; Sun, Z. Visualized study of thermochemistry assisted steam flooding to improve oil recovery in heavy oil reservoir with glass micromodels. *Fuel* **2018**, *218*, 118–126. [CrossRef]
4. Zhou, S.W.; Guo, L.J.; Li, Q.P. Multi-Source Multi-Thermal Fluid Generation System and Method. Patent CN106640007A, 5 October 2017.

5. Xu, J.L.; Kou, J.J.; Guo, L.J. Experimental study on oil-containing wastewater gasification in supercritical water in a continuous system. *Int. J. Hydrog. Energy* **2019**, *44*, 15871–15881. [CrossRef]
6. Sun, X.F.; Cai, J.M.; Li, X.Y. Experimental investigation of a novel method for heavy oil recovery using supercritical multi thermal fluid flooding. *Appl. Therm. Eng.* **2020**, *185*, 116–130.
7. Liu, Z.Y.; Li, Y.H.; Xu, L. Supercritical hydrothermal upgrading of oil sand: A review. *Chem. Ind. Eng. Prog.* **2018**, *37*, 4606–4615.
8. Ramazan, O.C.; Salim, S.S. Upgrading blends of microalgae feedstocks and heavy oils in supercritical water. *J. Supercrit. Fluids* **2018**, *133*, 674–682.
9. Li, Y.; Wang, S. Supercritical water oxidation for environmentally friendly treatment of organic wastes. In *Advanced Supercritical Fluids Technologies*; Pioro, I., Ed.; IntechOpen: Canada, 2019.
10. Liu, J.; Hu, N.; Fan, L.W. Simulation on the hydrogen oxidation reactor for supercritical H_2O/CO_2 working fluid. *J. Chem. Eng. Chin. Univ.* **2021**, *35*, 298–306.
11. Qin, Q.; Xia, X.B.; Li, S.B. Supercritical water oxidation and its application in radioactive waste treatment. *Ind. Water Treat.* **2022**, *42*, 38–45.
12. Liao, W.; Zhu, T.F.; Liao, C.H. Application of Supercritical Water Oxidation Technology in Energy Transformation. *Technol. Water Treat.* **2019**, *45*, 14–17.
13. Cocero, M.J.; Alonso, E.; Sanz, M.T.; Fdz-Polanco, F. Supercritical water oxidation process under energetically self-sufficient operation. *J. Supercrit. Fluids* **2002**, *24*, 37–46. [CrossRef]
14. Wang, Q.; Lv, Y.K.; Zhang, R. Supercritical Water Oxidation Kinetics of Textiles Dyeing Wastewater. *Adv. Mater. Res.* **2013**, *2203*, 634–638. [CrossRef]
15. Zan, Y.F.; Wang, S.Z.; Zhang, Q.M. Experimental Studies on the Supercritical Water Oxidation of Municipal Sludge and its Reaction Heat. *J. Chem. Eng. Chin. Univ.* **2006**, *03*, 379–384.
16. Wang, Y.Z.; Yu, H.; Sheng, J.P. Treatment of biochemical sludge from coal gasification wastewater by supercritical water oxidation process. *Chem. Eng.* **2015**, *43*, 11–15.
17. Xia, Q.Y.; Guo, W.M.; Shen, Z.M. Research on Supercritical Water Oxidation of Industrial Wastewater. *Saf. Environ. Eng.* **2014**, *21*, 78–83.
18. Ma, H.; Yang, Y.; Chen, Z. Numerical simulation of bitumen recovery via supercritical water injection with in-situ upgrading. *Fuel* **2022**, *313*, 122708. [CrossRef]
19. Xi, H.J.; Li, D.Y.; Zhang, L.C. Evaluation on Supercritical Steam Stimulation of Eastern Lukeqin Heavy Oil Wells. *Liaoning Chem. Ind.* **2013**, *42*, 1112–1114.
20. Wang, R. *Research on Supercritical Cyclic Steam Stimulation of Ultra-Deep Super Heavy Oil in East Area of Lukeqin, China*; University of Petroleum (East China): Qingdao, China, 2015.
21. Yin, Q.G.; Huo, H.B.; Hang, P.; Wang, C.; Wu, D.; Luo, Z. Effective Use of Technology Research and Application of Supercritical Steam on Deep Super-heavy Oil in Lukeqin Oilfield. *Oilfield Chem.* **2017**, *34*, 635–641.
22. Zhao, Q.Y.; Guo, L.J.; Huang, Z.J.; Chen, L.; Jin, H.; Wang, Y. Experimental Investigation on Enhanced Oil Recovery of Extra Heavy Oil by Supercritical Water Flooding. *Energy Fuels* **2018**, *32*, 1685–1692. [CrossRef]
23. Zhang, F.Y.; Pang, Z.X.; Wu, T.T.; Geng, Z.; Liao, H. Extra—Heavy Oil Hydrothermal Pyrolysis Experiment with Supercritical Steam. *Spec. Oil Gas Reserv.* **2019**, *26*, 130–135.
24. Zhao, Q.Y.; Guo, L.J.; Wang, Y.H.; Jin, H.; Chen, L.; Huang, Z. Enhanced Oil Recovery and in Situ Upgrading of Heavy Oil by Supercritical Water Injection. *Energy Fuels* **2019**, *34*, 360–367. [CrossRef]
25. Zhao, Q.Y.; Guo, L.J.; Wang, Y.C.; Huang, Z.; Chen, L.; Jin, H. Thermophysical Characteristics of Enhanced Extra-heavy Oil Recovery by Supercritical Water Flooding. *J. Eng. Thermophys.* **2020**, *41*, 635–642.
26. Wang, S.T.; Zhang, F.Y.; Liu, D.; Zhu, Q.; Ge, T.T. Physical—Numerical Simulation Integration of Super—Critical Steam—Flooding in Offshore Extra—Heavy Oil Reservoir. *Spec. Oil Gas Reserv.* **2020**, *27*, 138–144.
27. Zhang, F.Y.; Wang, S.T.; Liu, D.; Gao, Z.; Zhu, Q. Experimental study on the EOR mechanisms of developing extra heavy oil by supercritical steam. *Oil Drill. Prod. Technol.* **2020**, *42*, 242–246.
28. Han, S. *Mechanism and Application of Supercritical Cyclic steam for Heavy Oil Reservoirs*; Northeast Petroleum University: Daqing, China, 2020.
29. Zhao, Q.Y.; Wang, Y.C.; Zou, S.F.; Huang, Z.J.; Chen, L.; Guo, L.J. Numerical Study on Flow and Heat Transfer of Supercritical Water Injection in Wellbore. *J. Eng. Thermophys.* **2018**, *39*, 2213–2218.
30. Zhang, Z.Q. *Study on Flow Characteristics and Thermal Diffusion Law of Supercritical Steam*; China University of Petroleum (Beijing): Beijing, China, 2019.
31. Liu, C.T.; Zhang, W.; Yin, J.Q. Numerical simulation of phase change regularity of supercritical steam in vertical wellbores—By taking Gao 3624 block of Gaosheng oilfield as an example. *Pet. Geol. Eng.* **2020**, *34*, 106–111.
32. Pang, Z.X.; Gao, H.; Zhang, Z.Q.; Wang, L.T. Calculation and analysis of heat transfer characteristics during supercritical steam flow in vertical wellbore. *Pet. Geol. Recovery Effic.* **2021**, *28*, 111–118.
33. Lv, S.Y.; Li, Y.H.; Li, H.B.; Long, X.M. Predicting model of the physical property parameter of super-critical water injection wellbore in horizontal well of heavy oil reservoirs. *Pet. Geol. Oilfield Dev. Daqing* **2021**, *40*, 54–62.
34. Zou, S.F.; Guo, Z.B.; Wang, J.; Luo, Q.S. Numerical simulation of supercritical multi-component heat fluid injected into a wellbore. *Sci. Technol. Eng.* **2020**, *20*, 7663–7670.

35. Huang, Z.J.; Zhao, Q.Y.; Chen, L.; Miao, Y.; Wang, Y.C.; Jin, H.; Guo, L.J. Fundamentals of Enhanced Heavy Oil Recovery by Supercritical Multi-Component Thermal Fluid Flooding. *J. Eng. Thermophys.* **2022**, *43*, 974–981.
36. Sun, X.; Li, X.; Tan, X.; Zheng, W.; Zhu, G.; Cai, J.; Zhang, Y. Pyrolysis of heavy oil in supercritical multi-thermal fluid: An effective recovery agent for heavy oils. *J. Pet. Sci. Eng.* **2021**, *196*, 107784. [CrossRef]

Article

Laboratory Evaluation of the Plugging Performance of an Inorganic Profile Control Agent for Thermal Oil Recovery

Keyang Cheng [1,2,*], Yongjian Liu [1], Zhilin Qi [2], Jie Tian [2], Taotao Luo [2,*], Shaobin Hu [1] and Jun Li [2]

1 School of Petroleum Engineering, Northeast Petroleum University, Daqing 163318, China; lyj@nepu.edu.cn (Y.L.); hsbpaper@163.com (S.H.)
2 School of Petroleum and Natural Gas Engineering, Chongqing University of Science and Technology, Chongqing 401331, China; 2008008@cqust.edu.cn (Z.Q.); 2017026@cqust.edu.cn (J.T.); 2011927@cqust.edu.cn (J.L.)
* Correspondence: 2012905@cqust.edu.cn (K.C.); 2017039@cqust.edu.cn (T.L.)

Abstract: During the process of steam thermal recovery of heavy oil, steam channeling seriously affects the production and ultimate recovery. In this study, fly ash was used as the plugging agent, and then a series of plugging experiments based on the results of two-dimensional (2D) experiments were conducted to study the effect of plugging the steam breakthrough channels. The experimental results show that the inorganic particle plugging agent made from the fly ash had a good suspension stability, consolidation strength, and injection performance. Because of these characteristics, it was migrated farther in the formation with a high permeability than in the formation with a low permeability, and the plugging rate was greater than 99%. After steam injection, it had a good anti-flush ability and stable plugging performance in the formation. In terms of the oil displacement effect, oil recovery in the formation with a low permeability was effectively improved because of plugging. The results show that the inorganic particle plugging agent could effectively control the steam channeling and it improved the development effect of the heavy oil reservoir.

Keywords: fly ash; steam channeling; injection performance; plugging depth; heavy oil

Citation: Cheng, K.; Liu, Y.; Qi, Z.; Tian, J.; Luo, T.; Hu, S.; Li, J. Laboratory Evaluation of the Plugging Performance of an Inorganic Profile Control Agent for Thermal Oil Recovery. *Energies* **2022**, *15*, 5452. https://doi.org/10.3390/en15155452

Academic Editor: Daoyi Zhu

Received: 7 June 2022
Accepted: 20 July 2022
Published: 27 July 2022

1. Introduction

With the increasing demand for fossil energy, the development and utilization of heavy oil has attracted significant attention. Heavy oil is characterized by a high viscosity, high density, and high content of heavy components [1]. Thermal oil recovery is an effective method to develop heavy oil resources, including steam stimulation, steam flooding, in situ combustion, and steam assisted gravity flooding, of which steam stimulation and steam flooding are the most widely used [2–4]. However, steam channeling is the main problem regarding the steam thermal recovery of heavy oil owing to the heterogeneity of the formation and the difference of fluidity between the steam and heavy oil, which seriously affects the production and ultimate recovery of heavy oil [5]. The permeability is a key parameter to the fluid flow in porous media [6,7]. Therefore, it is crucial to weaken the negative impact of steam channeling.

Chemical plugging technology is an effective method to plug the steam breakthrough channels, which has the potential to greatly improve the development effect of heavy oil reservoirs [8]. At present, polymers, gels, particles, foams, resins, and microbials are widely used as plugging agents. Hu et al. [9] systematically investigated the EOR performance of foam flooding in different types of formations, and the results show that the flow resistance of the foam was promoted as the permeability increased. However, polymer gels, polymer microspheres, and expanded particles are not only expensive, but also cannot meet the technical requirements of the plugging strength needed in formations with a high permeability [10,11]. Foam plugging agents are limited by their thermal stability, and many low-temperature effective foam stabilizing agents cannot be used at high temperatures [12,13]. A short consolidation time and high cost limit the application of resin plugging agents [14]. The inorganic particle plugging agents are an ideal material owing to their high temperature and salt resistance, good stability, high compressive strength, and wide range of consolidation time adjustments [15,16].

Inorganic solid particles are mostly used as anti-channeling agents, such as cement and clay. The anti-channeling effect mostly depends on the particles' physical blockage. Although an anti-channeling agent of this kind has a good temperature resistance, it cannot reach the depth of formation because of the inferior injection performance, short plugging distance, and continuous fingering phenomenon that occurs after the steam flooding. After a long duration of high temperature aging, the increasing pores in porous media may shorten the plugging validity. In addition, they may cause greater damage due to their non-selective plugging [17]. Cement was first applied to plug steam breakthrough channels, but its application was limited by the short consolidation time and inability to realize deep profile controlling [18].

Inorganic solid particles, including fly ash, oily sludge, and scum, can be made into plugging agents. The plugging agent is composed of inorganic ultrafine solid particles, retarder, an adhesive-enhancing agent, and a consolidation agent. These solid particles have a strong compatibility with the formation. Their application also solves environmental problems to a certain extent, so they are widely used to plug the steam breakthrough channels in heavy oil reservoirs [19]. Fly ash and slag are similar to cement, and are hydraulic cementitious materials with a potential activity [20]. The temperature and activator are necessary conditions for activating the fly ash and slag. By adjusting the temperature and dosage of the activator, the solidification time of the cementation materials, such as fly ash and slag, can be controlled [21]. The solidified mechanism of fly ash is similar to the cement solidification process. In an alkaline environment, the mineral forms a variety of hydrates with a crystal structure around the micro particles through a series of chemical reactions and ion exchange in various ways. Under the action of molecular forces, the colloidal particles composed of micro particles and hydrates condense into a network structure, and then dehydrate to form a semi-rigid plate material [22]. Pan et al. [23] used industrial slag as the main agent to prepare a high-temperature plugging agent. The experimental results show that the plugging agent could effectively plug the steam breakthrough channels, and could then expand the sweep coefficient of the injected steam.

Here, we considered the bad adaptability of the traditional plugging agent. In this study, a series of sand-pack experiments based on the results of two-dimensional experiments were conducted to study the plugging effect of inorganic particle plugging agents in the steam breakthrough channels. By reducing the phenomenon of steam channeling through plugging the layer with a high permeability, we could adjust the steam injection profile, increase the steam sweep coefficient, and increase the ultimate recovery of heavy oil.

2. Characteristics of Steam Channeling between Two Wells

The 2D experiment of the steam stimulation intuitively displayed the process of steam channeling in two wells, and the experimental model is shown in Figures 1–3, namely the temperature distribution between the two wells during the steam stimulation, where the unit of temperature is °C. Figure 2 shows that the high temperature front of steam advanced along the layer with a high permeability, and the reservoir temperature gradually decreased with the increase in distance. With the increase in the stimulation cycles, the heating radius of the steam injection well increased, and the fluidity of the crude oil within the heating range increased. During the eighth steam stimulation, obvious steam channeling was formed between the two wells. Steam in the channeling layer had an ineffective flow, and the reservoir temperature and steam injection wellhead temperature exhibited little difference, as shown in Figure 3. Because there was positive rhythmic formation, the permeability difference between the upper and lower layers was 8.59, and the permeability of the lower layer was as high as 7.64 D. Steam easily passed through because of the low seepage resistance, so the phenomenon of steam channeling occurred in the lower layer during the process of steam stimulation.

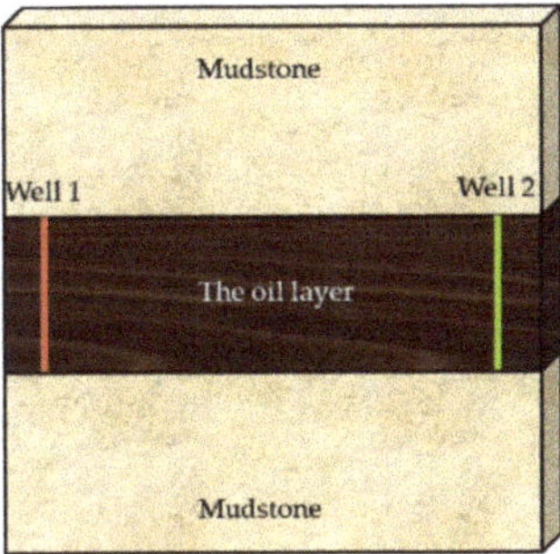

Figure 1. Schematic of the 2D experimental model.

Figure 2. Temperature distribution when the high temperature fronts made contact with each other.

Figure 3. Temperature distribution of steam channeling between the two wells.

It can be seen from the experimental results that steam is more likely to break through in the formation with a high permeability during the process of steam stimulation. Steam channeling leads to a decrease in the final recovery factor of heavy oil, greatly increases production cost, and seriously affects the development of heavy oil resources. Therefore, according to the results, a series of sand-pack experiments were conducted to study the adaptability of the inorganic particle plugging agent in steam breakthrough channels.

3. Experimental Section

3.1. Material

The oil samples were heavy oil from the Henan oilfield, and the viscosity of degassed crude oil at the reservoir temperature was 16,110–21,440 mPa·s. The salinity of the formation water was 1792 mg/L. The inorganic particle plugging agent made in the laboratory was composed of the fly ash, a suspension agent, a consolidation agent, and an auxiliary agent. The mass concentration of fly ash in the plugging agent was 25%, and the chemical composition of the fly ash is shown in Table 1. The sample of sand in the formation was used for filling the model of the sand-pack.

Table 1. Composition and content of the fly ash.

SiO_2(%)	Al_2O_3(%)	Fe_2O_3(%)	CaO(%)	MgO(%)	K_2O(%)	Na_2O(%)
55.63	18.38	6.75	4.93	1.72	2.49	1.09

3.2. Experimental Setup

The one-dimensional experiment system is shown in Figure 4. It consisted of an injection module, main module, data acquisition module, and produced liquid collection module. The injection module was composed of the liquid injection subsystem. The steam injection subsystem was mainly composed of a high-pressure precision injection pump, intermediate piston vessel, air compressor, and steam generator. The main module was mainly composed of the sand-pack and high temperature electric heating sleeve. The working pressure was adjustable from 0 to 70 MPa, and the working temperature was adjustable from 0 °C to 200 °C. The data acquisition module consisted of the temperature and pressure sensor, differential pressure sensor, and control cabinet. The produced liquid collection module was composed of a receiving container, back pressure valve, and measuring instrument. The sand-pack was 100 cm long with an inner diameter of 3.8 cm, and there were six pressure monitoring points including the inlet, as shown in Figure 5. The basic data of the sand-packed models is shown in Table 2.

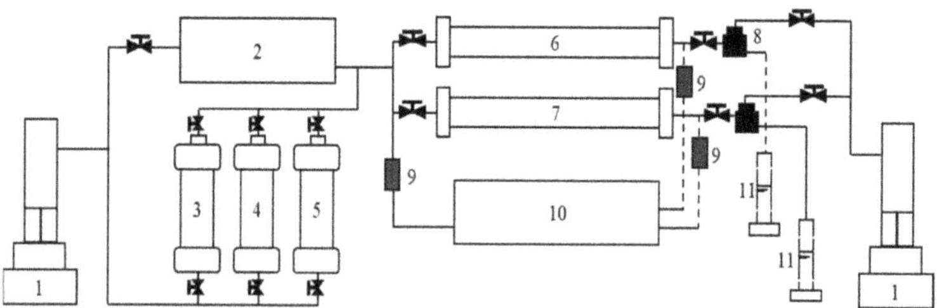

Figure 4. Schematic of the one-dimensional experiment. (1) ISCO pump; (2) steam generator; (3) formation water; (4) crude oil; (5) plugging agent; (6) high-permeability sand-pack; (7) low-permeability sand-pack; (8) back pressure regulator; (9) pressure sensor; (10) data acquisition system; (11) measuring cylinder.

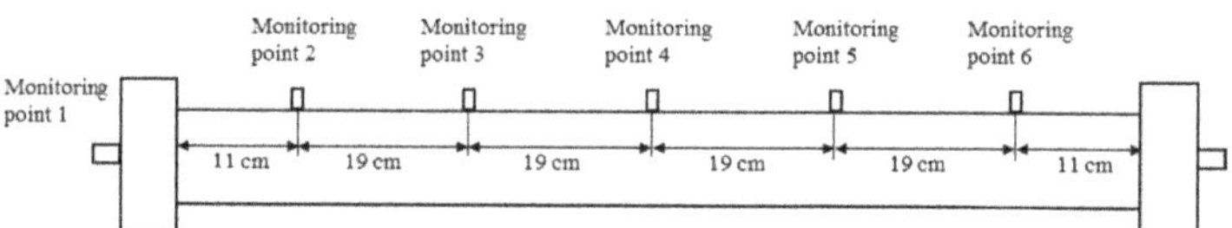

Figure 5. Distribution of the pressure monitoring points.

Table 2. The basic data of the sand-packed models.

Model Number	Diameter (cm)	Length (cm)	Pore Volume (mL)	Porosity (%)	Permeability (D)
1	3.8	100	302	27.32	1.15
2	3.8	100	359	32.46	7.23
3	3.8	100	316	28.54	0.99
4	3.8	100	373	33.67	7.68
5	3.8	100	298	26.85	1.09
6	3.8	100	355	31.98	7.16

3.3. Experimental Procedure

In order to study the plugging performance of the inorganic particle plugging agent in the formation with a high permeability, effectively solve the phenomenon of steam channeling in the process of steam stimulation, and improve the ultimate recovery of heavy oil, this study conducted suspension stability and consolidation strength tests, sand-pack plugged experiments, and one-dimensional physical simulation oil displacement experiments. The specific experimental process is as follows:

Suspension stability and consolidation strength test

The relative height sedimentation method was used for evaluating the performance of the inorganic particle plugging agent. After the inorganic particle plugging agent solution was stirred for 1 h, it was poured into a 25 mL centrifuge tube and placed in an oven at 80 °C for 6 h. The precipitation volume in the centrifuge tube was recorded over different stages. The change in plugging agent solution was observed for more than 6 h, including the initial set time, final set time, and consolidation process.

Plugging experiment

First, the plugging experiment was conducted to confirm the depth that the inorganic particle agent reached in a formation. After 1 pore volume (PV) of the plugging agent was injected into models No. 1 and No. 2 at a speed of 1 mL/min, the inlet and outlet valves were closed and placed in an oven at 80 °C to solidify for 12 h. The models were flooded with water at a speed of 1 mL/min after cooled, and the pressures of each monitoring point were recorded by the data acquisition system. Then, the steam flooding was conducted to evaluate the scour resistance ability. Then, 10 PV of steam was injected separately into models No. 1 and No. 2, and the inlet and outlet pressure were recorded. Finally, the plugging experiment was conducted in the parallel model, which was composed of models No. 3 and No. 4. Under the same experimental conditions, 0.5 PV of the inorganic particle plugging agent was injected, and then 0.05 PV of water was injected at a speed of 1 mL/min. The water was injected into this model at a speed of 1 mL/min, and the inlet and outlet pressure were recorded.

Oil displacement experiment

First, models No. 5 and No. 6 were saturated with oil, and the oil saturation of the models was calculated according to the volume of water production. Then, models No. 5 and No. 6 were connected in a parallel model, and steam was injected into this model until the water cut increased to 98%; meanwhile, the oil and liquid production were recorded. Then, 0.5 PV of the inorganic particle plugging agent was injected, and then 0.05 PV of water was injected at a speed of 1 mL/min. Finally, steam was injected into this model until the water cut increased to 98%; meanwhile, the oil and liquid production were recorded.

4. Experiment Results

4.1. Stability Evaluation

The main component of the inorganic particle plugging agent was fly ash, which was prepared by adding a suspension agent, consolidation agent, and auxiliary agent. The purpose of adding a suspension agent was to improve the suspension stability, and the role of the consolidation agent was to activate the fly ash and to generate hydration with the network structure; this hydration reaction rate was regulated by the auxiliary agent. The inorganic particle plugging agent solution was injected into the target formation in the form of a suspension, and its suspension stability, final set time, and consolidation strength were important indexes to evaluate the plugging effect. The smaller the volume of water separated from the inorganic particle plugging agent solution, the better the suspension performance. The final set time and consolidation strength were evaluated using the visual evaluation method. The consolidation time was determined by the strength of the inorganic particle plugging agent system.

As can be seen from Figure 6, with the increase in time, the precipitation volume of the plugging agent system in the centrifuge tube had an insignificant change. The precipitation volume decreased from 25 mL to 24.2 mL, with a change of 0.8 mL. After more than 4 h, the precipitated volume appeared to be stable, and the water extraction rate was 3.2%, indicating that the plugging agent solution had a good suspension stability.

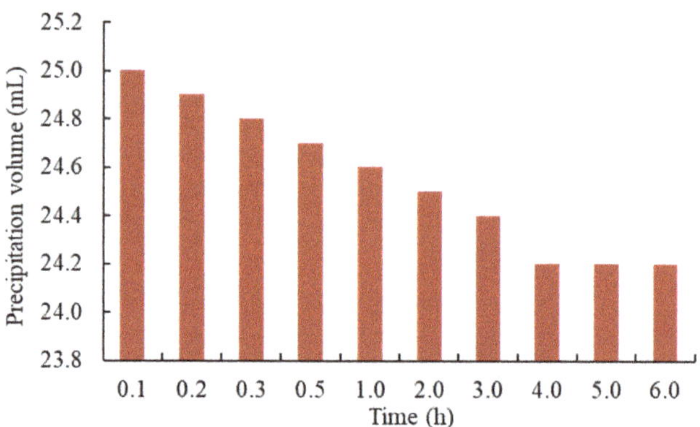

Figure 6. Precipitation volume of the plugging agent solution versus time.

Hydraulic compounds were generated from the plugging agent solution through some microscopic chemical reactions, and the reaction degree was macroscopically expressed as the strength change in the plugging agent. Table 3 shows that the initial setting time of the plugging agent solution was 6 h, with a relatively high fluidity. After 10 h, the formation of the plugging agent system was stable, and it had a relatively high strength. After more than 12 h, the plugging agent system had a rigid strength. The results show that the plugging agent solution had a good consolidation strength and could meet the requirements needed to plug larger pores.

Table 3. Consolidation time and consolidation strength.

Number	Time (h)	Appearance
1	3	Suspension, and high fluidity
2	6	Comparatively high fluidity, and beginning of consolidation
3	8	Medium fluidity, and unstable form
4	10	Stable form, and relatively high strength
5	11	Stable form, and high strength
6	12	Rigid strength, and the glass rod cannot enter
7	13	Rigid strength, and the glass rod cannot enter

4.2. Plugging Evaluation

The anti-channeling agent should not only have a strong temperature resistance, but it should also have an easy injection, in order to achieve plugging in the deep formation. Therefore, the plugging experiment was conducted to confirm the depth that the inorganic particle reached in the formation. Compared with the monitored pressure before the consolidation of the inorganic particle plugging agent, the plugging distance was determined by the changes in pressure at each monitoring point on the sand-pack model after solidification. Figures 7 and 8 show the changes in pressure at each monitoring point on the sand-packed model with an initial permeability of 1.15 D and 7.23 D, respectively. Each point in the curve is the ratio of pressure during water injection to stable pressure before injecting the plugging agent.

As can be seen from Figure 7, with the increase in the injection amount, the pressure ratio increased in monitoring points 1 to 5. Before monitoring point 3, the pressure ratio rose rapidly in the initial stage, and the curve fluctuated sharply after reaching the highest point, and finally tended to be basically stable. This phenomenon indicated that inorganic particle plugging agent had a good injection ability and could enter the middle and deep position of the sand-packed model. The hydraulic compound increased the seepage resistance of the model and caused a change in pressure. The pressure ratio at measuring point 6 was a relatively smooth straight line, indicating that the plugging agent could not reach the point. According to the calculation formula of the plugging rate, the greater the pressure difference, the greater the flow resistance of the formation, indicating that the inorganic particle plugging agent had a higher consolidation strength in the formation. The pressure difference between monitoring points 2 and 3 was large, indicating that the main plugging section of model No. 1 was between them.

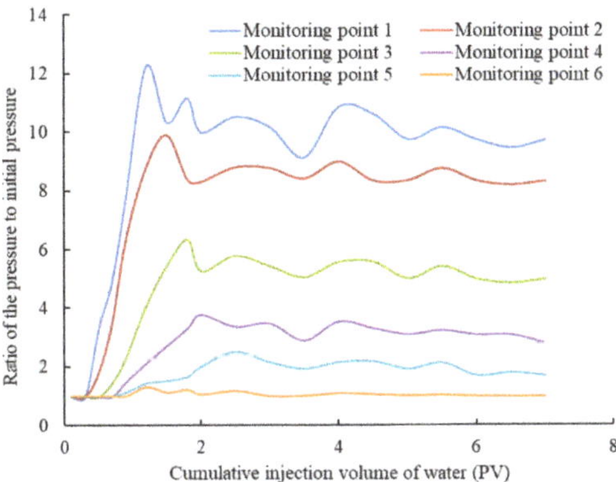

Figure 7. Pressure changes at each monitoring point for model No. 1 after plugging.

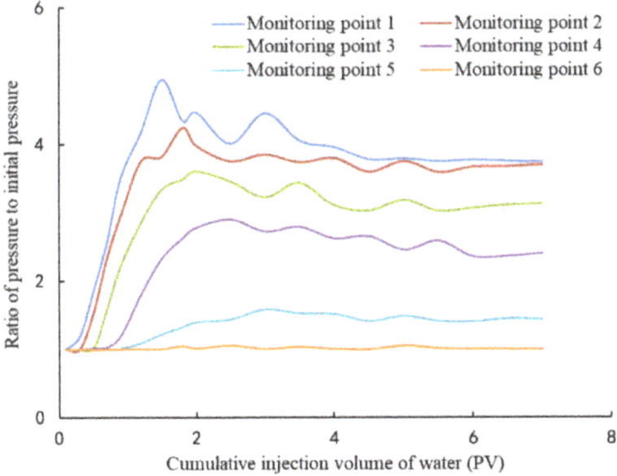

Figure 8. Pressure changes at each monitoring point for model No. 2 after plugging.

The difference between Figures 7 and 8 is the initial permeability of the sand-packed model. Figure 8 shows the plugging effect of the inorganic particle plugging agent in the formation with a high permeability. With the increase in injection volume, the pressure ratio increased in monitoring points 1 to 5. The main reason for the rapid rise in the pressure ratio before monitoring point 4 was that a large number of hydraulic compounds were generated at this position, which increased the seepage resistance and the pressure. It also indicates that the inorganic particle plugging agent had reached this position. The pressure ratio at measuring point 6 was a relatively smooth straight line, but the pressure ratios were all greater than 1, indicating that a small amount of plugging agent had reached the point. The pressure difference between measuring points 4 and 5 was large, indicating that the main plugging section of model No. 2 was between them.

Comparing the results of the two models, it can be seen that the inorganic particle plugging agent had a good injection performance and plugging effect, and could reach the middle and deeper part of the formation before consolidation. Compared with the formation with a low permeability, the plugging agent migrated farther in the formation with a high permeability. The plugging rate was greater than 99% in models No. 1 and No. 2.

Figure 9 shows the change in the displacement differential pressure with the volume of steam injection after the plugging agent consolidation. With the increase in the injection volume, the displacement pressure difference of models No. 1 and No. 2 were relatively stable with a small fluctuation range. After a continuous steam injection of 10 PV, the pressure difference of model No. 1 remained at 12.9 MPa and model No. 2 remained at 3.8 MPa; this indicates that the solidified plugging agent exhibited good adsorption and retention in the formation, excellent scouring resistance, and a stable plugging performance.

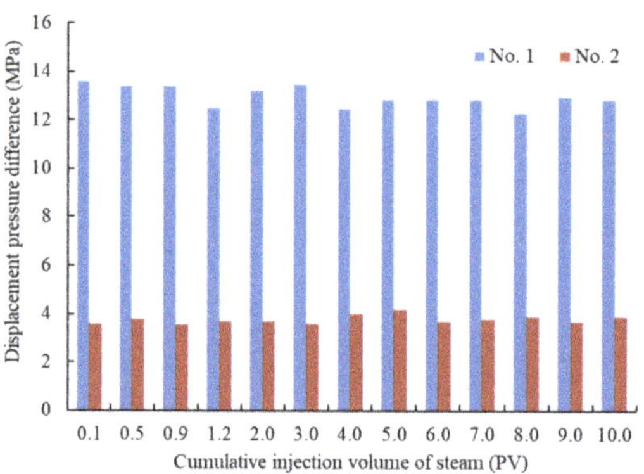

Figure 9. Changes in the displacement differential pressure.

4.3. Selective Plugging in Porous Media

As can be seen from Figure 10, the plugging rate of the inorganic particle plugging agent reached 99.18% for the formation with a high permeability, and only 10.1% for the formation with a low permeability. It can be seen that the plugging effect of the inorganic particle plugging agent had a selective plugging effect, the layer with a high permeability was effectively plugged, and a small amount of plugging agent entered the low permeability layer.

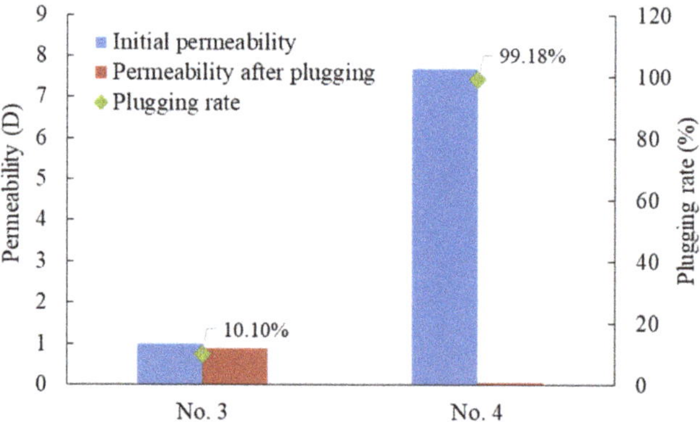

Figure 10. Plugging rate with different sand-packs in the parallel model.

4.4. The Displacement Effect

Table 4 shows the experimental results of the oil displacement in the parallel model. The fractional flow rate of the sand-packed model (No. 5) with a low permeability increased by 23%, while that of the sand-packed model (No. 6) with a high permeability was effectively reduced. The recovery factor of models No. 5 and No. 6 was improved by 24.34% and 4.45%, respectively. It could be seen that the inorganic particle plugging agent could effectively plug the high permeability formation, adjust the steam injection profile, expand the scope of the steam sweep, and improve the final recovery of the heavy oil reservoir.

Table 4. Experimental data of the oil displacement.

Model Number	Fractional Flow Rate (%)	Pre-Plug Recovery Factor (%)	Fractional Flow Rate after Plugging (%)	Recovery Factor after Plugging (%)	Enhanced Oil Recovery (%)
5	25.00	31.28	48.00	55.62	24.34
6	75.00	60.33	52.00	64.78	4.45

5. Conclusions

In this study, through the performance evaluation of the inorganic particle plugging agent and laboratory experiments, the following conclusions were obtained.

(1) The inorganic particle plugging agent made from the fly ash had a good suspension stability and consolidation strength. The water extraction rate was 3.2%, the consolidation time was 12 h, and the consolidation strength was rigid. It met the requirements for plugging large pores.

(2) The inorganic particle plugging agent made from the fly ash had a good injection performance and plugging effect. The plugging depth was longer in the high-permeability formation, the plugging rate was greater than 99%, and the scour resistance was strong. It had the characteristic of "plugging larger pores but not plugging small" and showed little damage to the low permeability formation.

(3) The inorganic particle plugging agent made from the fly ash could effectively plug the steam breakthrough channels, which could improve the steam injection profile and the thermal recovery of the heavy oil reservoir.

Author Contributions: K.C., Z.Q. and Y.L. conceived and designed the experiments; K.C., T.L. and J.L. performed the experiments; K.C., J.T. and S.H. analyzed the data; K.C. and T.L. wrote the paper. All authors have read and agreed to the published version of the manuscript.

Funding: This research was funded by the Science and Technology Research Project of the Chongqing Municipal Education Commission of China (grant no. KJQN201901540, no. KJQN201901542).

Institutional Review Board Statement: Not applicable.

Informed Consent Statement: Not applicable.

Data Availability Statement: The data that support the findings of this study are available from the corresponding author.

Conflicts of Interest: The authors declare no conflict of interest.

References

1. Im, S.I.; Shin, S.; Park, J.W.; Yoon, H.J.; Go, K.S.; Nho, N.S.; Lee, K.B. Selective separation of solvent from deasphalted oil using CO_2 for heavy oil upgrading process based on solvent deasphalting. *Chem. Eng. J.* **2018**, *331*, 389–394. [CrossRef]
2. Dong, X.; Liu, H.; Chen, Z.; Wu, K.; Lu, N.; Zhang, Q. Enhanced oil recovery techniques for heavy oil and oilsands reservoirs after steam injection. *Appl. Energy* **2019**, *239*, 1190–1211. [CrossRef]
3. Guo, K.; Li, H.; Yu, Z. In-situ heavy and extra-heavy oil recovery: A review. *Fuel* **2016**, *185*, 886–902. [CrossRef]
4. Zhao, S.; Pu, W.-F.; Su, L.; Shang, C.; Song, Y.; Li, W.; Het, H.-Z.; Liu, Y.-G.; Liu, Z.-Z. Properties, combustion behavior, and kinetic triplets of coke produced by low-temperature oxidation and pyrolysis: Implications for heavy oil in-situ combustion. *Pet. Sci.* **2021**, *18*, 1483–1491. [CrossRef]
5. Liu, H.; Wang, Y.; Zheng, A.; Sun, X.; Dong, X.; Li, D.; Zhang, Q. Experimental investigation on improving steam sweep efficiency by novel particles in heavy oil reservoirs. *J. Pet. Sci. Eng.* **2020**, *193*, 107429. [CrossRef]
6. Xiao, B.; Wang, W.; Zhang, X.; Long, G.; Fan, J.; Chen, H.; Deng, L. A novel fractal solution for permeability and Kozeny-Carman constant of fibrous porous media made up of solid particles and porous fibers. *Powder Technol.* **2019**, *349*, 92–98. [CrossRef]
7. Liang, M.; Fu, C.; Xiao, B.; Luo, L.; Wang, Z. A fractal study for the effective electrolyte diffusion through charged porous media. *Int. J. Heat Mass Transf.* **2019**, *137*, 365–371. [CrossRef]

8. Li, X.; Yue, X.; Yue, T.; Yan, R.; Olaleye, O.E.; Shao, M. Novel chemical flood combination of CSA particles and strong emulsifying surfactant in heterogeneous reservoirs. *Colloids Surf. A Physicochem. Eng. Asp.* **2021**, *625*, 126838. [CrossRef]
9. Hu, L.Z.; Sun, L.; Zhao, J.Z.; Wei, P.; Pu, W.F. Influence of formation heterogeneity on foam flooding performance using 2D and 3D models: An experimental study. *Pet. Sci.* **2020**, *17*, 734–748. [CrossRef]
10. Liu, D.; Shi, X.; Zhong, X.; Zhao, H.; Pei, C.; Zhu, T.; Zhang, F.; Shao, M.; Huo, G. Properties and plugging behaviors of smectite-superfine cement dispersion using as water shutoff in heavy oil reservoir. *Appl. Clay Sci.* **2017**, *147*, 160–167. [CrossRef]
11. AlSofi, A.M.; Blunt, M.J. Polymer flooding design and optimization under economic uncertainty. *J. Pet. Sci. Eng.* **2014**, *124*, 46–59. [CrossRef]
12. Eftekhari, A.A.; Krastev, R.; Farajzadeh, R. Foam stabilized by fly-ash nanoparticles for enhancing oil recovery. In Proceedings of the SPE Kuwait Oil and Gas Show and Conference, Mishref, Kuwait, 11–14 October 2015; SPE-175382-MS. [CrossRef]
13. Seright, R.; Brattekas, B. Water shutoff and conformance improvement: An introduction. *Pet. Sci.* **2021**, *18*, 450–478. [CrossRef]
14. Zhao, G.; Dai, C.; Chen, A.; Yan, Z.; Zhao, M. Experimental study and application of gels formed by nonionic polyacrylamide and phenolic resin for in-depth profile control. *J. Pet. Sci. Eng.* **2015**, *135*, 552–560. [CrossRef]
15. Assi, L.N.; Deaver, E.; Elbatanouny, M.K.; Ziehl, P. Investigation of early compressive strength of fly ash-based geopolymer concrete. *Constr. Build. Mater.* **2016**, *112*, 807–815. [CrossRef]
16. Naghizadeh, A.; Ekolu, S.O. Method for comprehensive mix design of fly ash geopolymer mortars. *Constr. Build. Mater.* **2019**, *202*, 704–717. [CrossRef]
17. You, Q.; Wang, H.; Dai, C.L.; Liu, Y.; Fang, J.; Zhao, G.; Zhao, M.; Wang, K. Case study on in-depth blocking for thermal recovery by steam stimulation in heavy oil reservoirs. *SPE* **2015**, *174596*, 360–376. [CrossRef]
18. Akbari, S.; Taghavi, S.M. Fluid experiments on the dump bailing method in the plug and abandonment of oil and gas wells. *J. Pet. Sci. Eng.* **2021**, *205*, 108920. [CrossRef]
19. Raffa, P.; Broekhuis, A.A.; Picchioni, F. Polymeric surfactants for enhanced oil recovery: A review. *J. Pet. Sci. Eng.* **2016**, *145*, 723–733. [CrossRef]
20. Grant Norton, M.; Provis, J.L. 1000 at 1000: Geopolymer technology—The Current State of the Art. *J. Mater. Sci.* **2020**, *55*, 13487–13489. [CrossRef]
21. Yao, X.; Yang, T.; Zhang, Z. Fly ash-based geopolymers: Effect of slag addition on efflorescence. *J. Wuhan Univ. Technol. Mater. Sci. Ed.* **2016**, *31*, 689–694. [CrossRef]
22. Duxson, P.; Fernández-Jiménez, A.; Provis, J.L.; Lukey, G.C.; Palomo, A.; Van Deventer, J.S.J. Geopolymer technology: The current state of the art. *J. Mater. Sci.* **2007**, *42*, 2917–2933. [CrossRef]
23. Pan, J.; Deng, W.; Wang, Z. Studies and application of slag-blocking agent for steam-injection reservoir. *SPE* **2006**, *104426*, 1036–1041. [CrossRef]

Article

Study on Unblocking and Permeability Enhancement Technology with Rotary Water Jet for Low Recharge Efficiency Wells in Sandstone Geothermal Reservoirs

Chao Yu [1], Tian Tian [2], Chengyu Hui [1], Haochen Huang [1] and Yiqun Zhang [1,*]

1 State Key Laboratory of Petroleum Resources and Prospecting, China University of Petroleum-Beijing, Beijing 102249, China
2 Engineering Research Institute, Sinopec Shanghai Offshore Oil & Gas Company, Shanghai 200120, China
* Correspondence: zhangyq@cup.edu.cn

Abstract: In China, sandstone geothermal reservoirs are large in scale and widely distributed, but their exploitation is hindered by low recharge efficiency. In this paper, an unblocking and permeability enhancement technology using a rotary water jet for low recharge efficiency wells in sandstone geothermal reservoirs is proposed to solve this problem. This paper presents a series of studies about the proposed technology, including experiments, simulation and field application. Firstly, an experiment was carried out to verify the scale removal effect of a high-pressure water jet on the inner wall of the screen tube and its impact on sandstone. Secondly, the numerical models of the rotary jet flow field in the wellbore were established by ANSYS Fluent to study the influence of parameters. Finally, based on the simulation and experiment results, a rotary jet tool applicable to unblocking and descaling low-efficiency wells was designed, and a field application for low-efficiency wells in sandstone thermal reservoirs was conducted. The study results show that the unblocking and permeability enhancement technology using a rotary water jet is effective in removing the blockages and improving the permeability near the well. In conclusion, the presented technology can solve the problem of low efficiency during the reinjection of cooled thermal waters back into sandstone geothermal reservoirs and has great effectiveness in field application.

Keywords: sandstone reservoir; rotary jet; recharge efficiency; unblocking; permeability enhancement

Citation: Yu, C.; Tian, T.; Hui, C.; Huang, H.; Zhang, Y. Study on Unblocking and Permeability Enhancement Technology with Rotary Water Jet for Low Recharge Efficiency Wells in Sandstone Geothermal Reservoirs. *Energies* **2022**, *15*, 9407. https://doi.org/10.3390/en15249407

Academic Editor: Reza Rezaee

Received: 2 November 2022
Accepted: 6 December 2022
Published: 12 December 2022

1. Introduction

Geothermal energy is a non-polluting renewable energy source compared to traditional fossil energy sources, which has the advantages of local access, stable operation and clean environmental protection [1,2]. Geothermal resources mainly comprise hot dry rock and hydrothermal resources, the latter requiring less technical requirements and being more economical to develop [3–5]. China's sandstone geothermal resources have an abundant, widely distributed and high development potential, which has gradually increased in development scale in recent years [6]. Utilization of geothermal fluids is a common method of hydrothermal resource development, but this method may result in different potential geological risks, such as lowering the groundwater table and ground subsidence [7,8]. Reinjection of cooled geothermal water back into the reservoir is one of the solutions [9,10]. Suspended particles in the fluid accumulate in the seepage channels during reinjection and form blockages near the wellbore, reducing the permeability of the near-well zone and ultimately causing a decrease in reinjection volume or an increase in rejection pressure in the development well [11].

Physical, chemical or biological plugs can occur during geothermal tailwater reinjection in different combinations. Physical plugs during geothermal tailwater reinjection are mainly blockages caused by suspended matter, particle transport and clay swelling. When the rechargeable geothermal reservoir is characterized by high ground pressure, loose cementation, high clay content, low permeability and poor water circulation, these physical plugs are likely to occur [12,13]. During the reuse of geothermal water, the high concentration of chemical ions such as chloride ion and sulfate ion in the water can corrode the metal casing or screen pipe and produce insoluble substances during the reinjection process to form a chemical plug in the underground reservoir [14,15]. Microbial blockage of recharge reservoirs involves both microbiological and biochemical mechanisms of action, including colonization by the bacteria themselves and extracellular polymers, and may result in permanent blockage [16,17].

Reasonable development and utilization of geothermal energy are of great significance for reducing carbon emissions and optimizing the energy structure. The effective development and utilization of hydrothermal geothermal resources cannot be achieved without the development of unblocking technology. The velocity distribution of the swirling jet is helical, which can significantly enhance the turbulence pulsation and is conducive to flushing the attached blockage particles [18]. Cavitation occurs in the process of plug removal by rotary jet, which can clean the blockage in the seepage hole throat [19]. Based on the reservoir characteristics of sandstone thermal reservoirs and the mechanism of the rotary jet [20–22], this paper proposes a rotary jet unblocking technology applicable to low-efficiency wells [23,24]. Through laboratory experiments and numerical simulations, the author verified the effectiveness of high-pressure jet descaling and unblocking and proposed construction recommendations in sandstone geothermal reservoir infiltration enhancement operations. After field operations, the reinjection level and well conditions of the sandstone geothermal reservoir inefficient well were effectively improved. This paper provides a theoretical basis and experimental reference for the unblocking and permeability enhancement of sandstone geothermal reservoirs based on rotary jet technology and provides reference and guidance for the application and promotion of rotary jet unblocking technology in sandstone geothermal reservoirs.

2. Experiment of the Jet Pressure Wave Transmission Efficiency

In this paper, the high-pressure water jet was applied for plug removal and permeability enhancement. An experiment on the transmission efficiency of jet pressure waves was carried out to verify the scale removal effect of a high-pressure water jet on the inner wall of the screen tube and its impact on sandstone. In order to realize this experiment, different types of sandstone rock samples were attached to the outer wall of the screen tube section and were washed through a high-pressure water jet, which is used to simulate the plug removal and jetting operation in practical application.

2.1. Experiment Principle

Previous scholars have conducted many studies on the plug removal principle of sandstone reservoirs and jet impact pressure [25–27]. In this paper, an experiment was carried out to obtain high-pressure water jet pressure wave transmission efficiency for screen tubes taken from low-efficiency reinjection wells in a sandstone thermal reservoir. The layout of the experiment device is shown in Figure 1.

As shown in Figure 1, the sandstone rock sample should be cut into an arc along the outer wall of the screen tube, and the sample and the screen tube should be fitted on the experiment platform before the experiment. The jet spray gun is placed inside the cut screen tube, and the inner wall of the screen tube is continuously flushed for a period of time by a high-pressure jet according to a certain pressure and volume flow. After jet flushing, the descaling effect of the water jet on the inner wall of the screen tube, the erosion and crushing effect on sandstone samples will be observed. Thus, it can be judged whether the high-pressure water jet can continue to erode and break the sandstone samples outside the screen tube after passing through the screen tube wall and maintaining a certain jet energy.

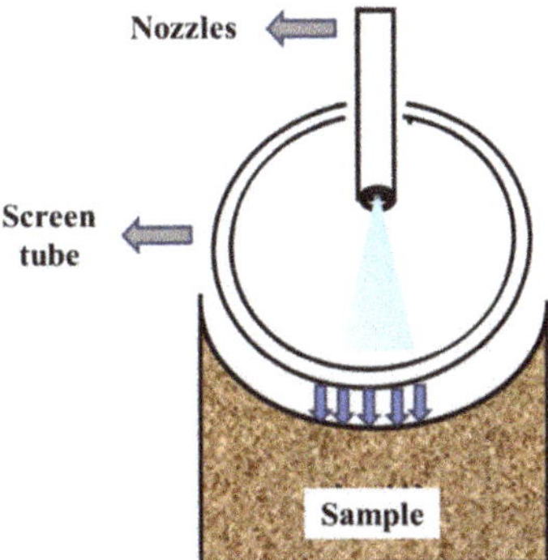

Figure 1. Experiment principal diagram.

2.2. Experiment Equipment and Materials

This experiment adopted a jet spray gun, whose structure is shown in Figure 2. In this experiment, the spray gun was equipped with three nozzles of 4.0 mm diameter. During the experiment, the screen tube could be sprayed at a fixed point, and the efficiency of the jet passing through the screen tube was studied. The experiment pump is a skid-mounted high-pressure three-cylinder plunger pump with a maximum volume flow of 0.7 m^3/min and a maximum pump pressure of 70 MPa.

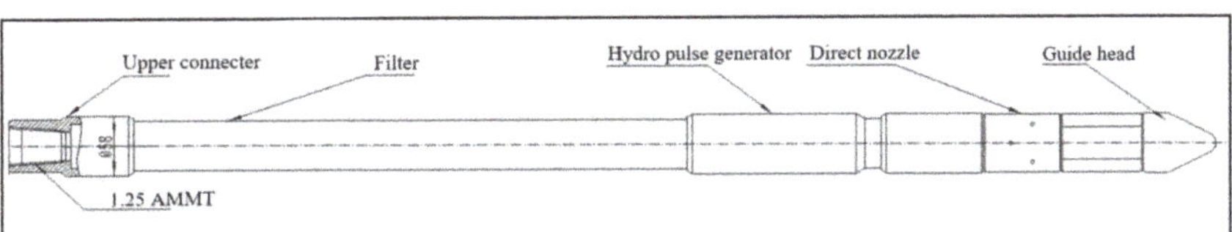

Figure 2. Schematic diagram of jet tool structure.

The screen tube used in the experiment is the wire wound screen tube used in the geothermal reinjection well of Guantao Formation in Lankao County, Henan Province. The screen tube had serious rust inside the base tube, and there was a blockage between the screen and the wire wound. Before the experiment, the reasons for the blockage of the screen tube were analyzed. As we know, the geothermal water of the Guantao Formation contains more minerals, and minerals in the geothermal water may be separated when the occurrence environment (temperature, pressure, PH value, etc.) of the local layer water changes, which form blockages and block the screen tubes and pore channels. Additionally, in the long-term injection production process, the sand production of the reservoir and the insoluble particles in the reinjection fluid also form sediment in the screen tube to block the screen tube channel, which is also the main reason for screen tube blockage. In addition, the screen tube can also be blocked by rust and sediment blockage formed by inorganic scale due to screen tube corrosion. The screen tube and inner wall used in the experiment are shown in Figure 3.

In the experiment of the jet pressure wave transmission efficiency, a yellow sandstone sample with a specification of 100 × 100 × 100 mm was adopted as the experiment rock sample. The physical parameters of the yellow sandstone type are shown in Table 1. In this experiment, the surface of the yellow sandstone rock sample was cut into an arc matching the outer wall of the screen tube to simulate the working conditions during downhole operation, which can represent the sandstone outside the well wall during the blockage removal operation to some extent. After the jet experiment, we observed whether the yellow sandstone was broken and eroded under the action of the jet to judge the efficiency of the water jet passing through the screen tube. The sandstone sample is shown in Figure 4.

Figure 3. Experiment screen tube and internal conditions.

Table 1. Physical parameters of yellow sandstone for experiment.

Parameter	Unit	Value
Density	g/cm^3	2.52
Uniaxial compressive strength	MPa	99.771
Elastic modulus	GPa	24.2
Poisson's ratio	/	0.182
Permeability	mD	0.0184
Porosity	%	4.33

Figure 4. The yellow sandstone samples for experiment.

2.3. Experiment Scheme and Procedures

The purpose of this experiment was mainly to study whether the high-pressure water jet can effectively remove the scaling and blocking state of the screen tube. After the jet passes through the screen tube, it can achieve a certain erosion and crushing effect on the sandstone rock sample on the outer wall of the screen tube and provides a reference and basis for the on-site rotary jet unblocking operation. For this reason, an experimental scheme for the experiment of jet pressure wave transmission efficiency was designed in this section. In order to make the experiment results as close as possible to the working conditions of the on-site unblocking operation, the experiment pump pressure was set to 26 MPa, and the spray distance was set to 5.0 mm. The experiment device is shown in Figure 5.

Figure 5. Experiment device fixed.

The specific steps for the experiment are as follows:

(1) Connect the jet tools and pipelines to the skid-mounted pump and then start the pump. Run under low pressure for a period of time to test the connectivity and sealing of the experimental device, and stop the pump after the test result is stable.

(2) Cut the screen tube into sections with a length of 0.5 m, and then fix the tube sections and sandstone rock sample on the experiment platform according to the designed experiment plan. At the same time, the screen tube shall be well supported to prevent the damage of sandstone samples caused by the self-weight of the screen tube from interfering with the test results.

(3) Connect the jet tools to the centralizer, place it in the screen tube and fix it. Fix the high-pressure flexible pipe on the experiment platform to prevent shaking during the experiment.

(4) Start the skid-mounted pump. Increase the pump speed gradually when the pressure of the skid-mounted pump is stable, then raise the pump pressure to the target pressure. Record the experimental data (volume flow, pump pressure and injection time), and evaluate the injection effect during the jet experiment

(5) After the experiment, reduce the speed and gear of the skid-mounted pump gradually and stop the pump when the pressure is stable. Then remove the pipeline and screen tube, and return the test device to its original position.

2.4. Experiment Results and Analysis

During the experiment, the pump pressure was maintained at about 26 MPa, and the volume flow was maintained at 0.56 m^3/min. The spraying effect of the high-pressure jet during the experiment and the scale removal results of the high-pressure jet are shown in Figure 6.

In the process of the experiment, a large amount of water was sprayed out through the screen tube within a short time after the jet started and the pump pressure was stable. During the experiment, the high-pressure water jet could keep on spraying to the highest position of about 2 m after it quickly passed through the screen tube, the internal screen and the winding wire. The experiment lasted for 5 min.

Figure 6. Experiment flushing process.

After the experiment, the jet spray gun was taken out, and the situation of the inner wall of the screen tube after continuous flushing by a high-pressure jet was observed. The results are shown in Figure 7. It can be seen that the rust of the inner wall of the screen tube after the experiment has been significantly improved under the action of the jet, and the scaling and blockage in the inner base tube hole have also been cleaned under the action of the jet. However, the high-pressure water jet did not damage the screen between the base pipe and the winding wire, and the clogging state of the screen mesh was also significantly improved. From the flushing results of the screen tube, it can be concluded that the high-pressure water jet can produce a more effective cleaning and descaling effect on the screen tube and the internal screen without damaging them, and after passing through the screen, the energy of the water jet can still be kept at a high level. The internal condition of the screen pipe after flushing is shown in Figure 7.

Figure 7. Interior of the screen tube and the screen after the experiment.

The screen tube and sandstone sample were separated after the jet experiment. The surface morphology of the yellow sandstone sample after jet spraying is shown in Figure 8. It is obvious that after continuous washing with a high-pressure water jet, there are many erosions shallow pits on the surface of the sandstone sample at the corresponding position of the jet, and the corners are also broken under the action of the jet. It can be seen that high-pressure jet has played a certain role in the erosion and destruction of sandstone samples after passing through the screen tube. In order to further study the erosion effect of a high-pressure water jet on sandstone, the yellow sandstone samples before and after the experiment were scanned through a three-dimensional surface appearance instrument.

Figure 8. Post-experimental yellow sandstone rock sample.

The three-dimensional surface appearance instrument used was provided by our laboratory, which can scan the strength and surface height of the target rock sample without damaging the rock sample. The appearance instrument scanning device is shown in Figure 9.

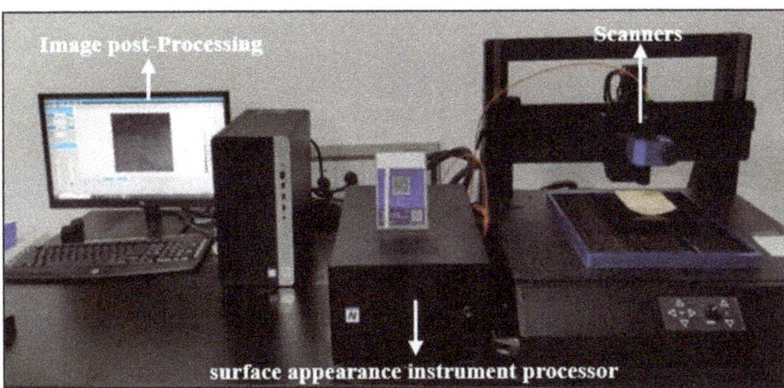

Figure 9. Three-dimensional surface appearance instrument.

The yellow sandstone samples before and after the experiment were scanned, and the scanning results are shown in Figure 10. By comparing the scanning results before and after the experiment, from the perspective of the scanning results of surface height, it can be seen that the erosion of shallow pits on the surface of the yellow sandstone rock sample after the flushing is also reflected in the scanning results. By comparing the surface strength of the yellow sandstone samples before and after the experiment, it can be seen that the overall strength of the sandstone sample surface also decreases after the high-pressure water jet flushing.

Based on the above results, it can be seen that a high-pressure water jet can play a certain role in removing the rust and scale on the inner wall of the screen pipe. The high-pressure jet can effectively pass through the screen tube wall, causing certain erosion, crushing and loosening effects on the sandstone samples on the outer wall without destroying the screen.

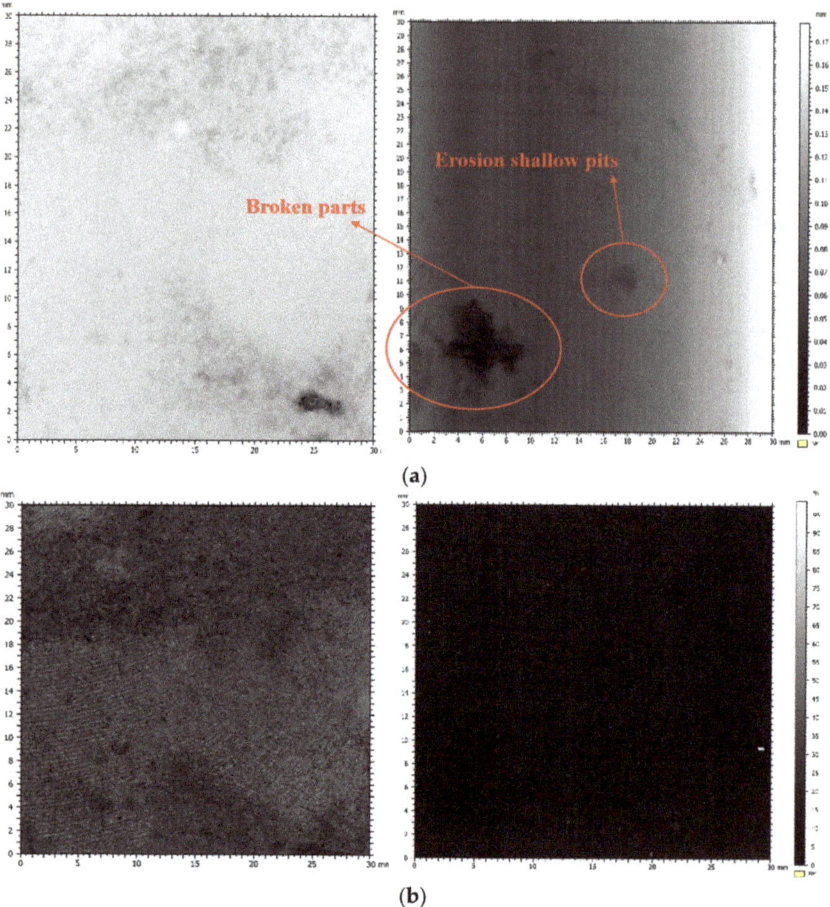

Figure 10. The scanning results of the three-dimensional appearance instrument before and after the experiment should be listed as follows: (**a**) the surface height scanning results before and after washing (**left**: before experiment; **right**: after experiment); (**b**) the surface intensity scanning results before and after washing (**left**: before experiment; **right**: after experiment).

3. Simulation Study of Rotary Jet Wellbore Flow Field

3.1. Model Description

In the process of unblocking and scale removal by rotary jet, the effect of rotary jet on the wellbore directly affects the final effect of unblocking and scale removal, and controlling the spray distance and volume flow of rotary jet tools plays an important role in the effect of the jet. In order to study the flow field characteristics and flushing range of the rotary jet in the wellbore, the flow field of the rotary jet in the wellbore was simulated using Fluent. The Eulerian multiphase model was selected, which allowed multiple independent but interacting phases to be modeled. The geometric model (Figure 11) is divided into casing area, rotation area, nozzle model and tubing area. The main setup area of the model is shown below, where the nozzle arrangement includes 1 forward nozzle, 2 inclined nozzles and 4 lateral nozzles. The model uses a 90 mm smooth cylinder to represent the wellbore, and the maximum outer diameter of the spray gun body is 30 mm. During the numerical simulation, the smooth cylinder used to simulate the wellbore remains stationary, the rotary tool rotates in the wellbore at a predetermined speed, and a rotary jet is formed in the wellbore to act on the inner wall of the wellbore. The diameter of nozzles is 3 mm, the volume flow is 1 m^3/min and the rotational speed is 100 rpm. In this study, we set the nozzle face on the tool as the velocity inlet, and the outlet is the annular section between the well wall and the tool tubing column, which is also set as the pressure outlet, and the well wall face is set as the wall type.

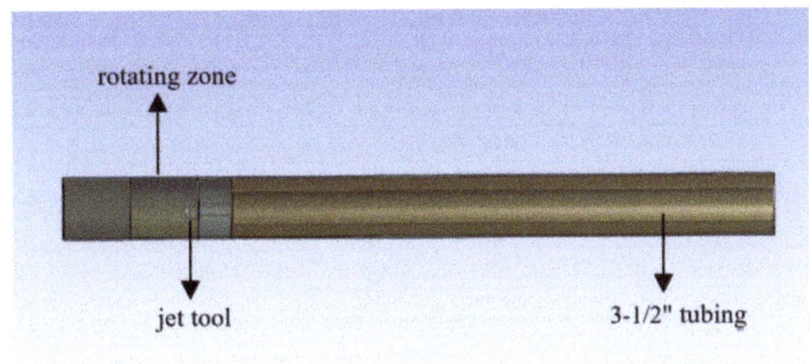

Figure 11. The diagram of the established model.

3.2. Governing Equations

In this study, the VOF model was used for the multiphase flow model in order to approximate the flow characteristics of the rotating jet in the wellbore under realistic conditions. The VOF model is a surface tracking method under a fixed Eulerian grid. In the VOF model, each fluid phase shares a common set of momentum equations, and each phase interface is tracked by introducing this variable of phase volume fraction. When we assume that the volume fraction of the q-th phase in the VOF model is α_q, then we can derive three cases: ① $\alpha_q = 0$, at which time we can know that there is no q-th phase fluid in the model cell; ② $0 < \alpha_q < 1$, at which time we can assume that the model cell contains multiphase fluid including qth phase; and ③ $\alpha_q = 1$, at which time we can assume that the model cell is filled with q-th phase fluid. The continuity equation and momentum equation of the VOF multiphase flow model are as follows.

$$\frac{\partial \alpha q}{\partial t} + \vec{v} \cdot \nabla a_q = \frac{S_{\alpha q}}{\rho_q} \tag{1}$$

In the continuity equation, $S_{\alpha q}$ is the mass source term, which is zero by default, and the volume fraction of each phase in the model is constrained to behave as follows:

$$\sum_{q=1}^{n} \alpha_q = 1 \tag{2}$$

$$\frac{\partial}{\partial t}\left(\rho \vec{v}\right) + \nabla\left(\rho \vec{v} \vec{v}\right) = \nabla p + \nabla\left[\mu\left(\nabla \vec{v} + \nabla \vec{v^T}\right)\right] + \rho \vec{g} + \vec{F} \tag{3}$$

where v denotes the fluid velocity, m/s; ρ denotes the density of the corresponding fluid, kg/m^3; $\vec{g}$ denotes the acceleration of gravity, m/s^2; μ denotes the viscosity of the phase fluid, N/m^3; and $\vec{F}$ denotes the interfacial force source term, N·m/s^2.

In this study, the realizable k-ε model was used for the calculation. The control equations of the k-ε model mainly include the k transport equation and the ε transport equation.

$$\frac{\partial}{\partial t}(\rho k) + \frac{\partial}{\partial x_j}\left[\left(\mu + \frac{\mu_t}{\sigma_k}\right)\frac{\partial k}{\partial x_j}\right] + G_k + G_b - \rho\varepsilon - Y_M + S_k \tag{4}$$

where k denotes the turbulent energy; μ denotes the viscosity of the fluid, N/m^3; G_k denotes the generation term of the turbulent energy k caused by the mean velocity gradient; σ_k denotes the Prandtl number corresponding to the turbulent energy k; and S_k denotes the source term.

$$\frac{\partial}{\partial t}(\rho\varepsilon) + \frac{\partial}{\partial x_j}(\rho\varepsilon\mu_j) = \frac{\partial}{\partial x_j}\left[\left(\mu + \frac{\mu_t}{\sigma_\varepsilon}\right)\frac{\partial \varepsilon}{\partial x_j}\right] + \rho C_1 S_\varepsilon - \rho C_2 \frac{\varepsilon^2}{k + \sqrt{v\varepsilon}} + C_{1\varepsilon}\frac{\varepsilon}{k}C_{3k}G_b + S_\varepsilon \quad (5)$$

$$C_1 = max\left[0.43, \frac{\eta}{\eta + 5}\right] \quad (6)$$

$$\eta = S\frac{k}{\varepsilon} \quad (7)$$

$$S = \sqrt{2S_{ij}S_{ij}} \quad (8)$$

where ε denotes the turbulent dissipation rate; σ_ε denotes the Prandtl number corresponding to the turbulent energy ε; C_1, C_2, $C_{1\varepsilon}$ and C_{3k} denote the coefficients in the equation; and S_ε denotes the source term.

3.3. Meshing Schemes

In the process of numerical simulation, the quality of grid division is very important to the results of numerical simulation. The finer the grid division, the higher the accuracy of the calculation results, but correspondingly, more calculation time is required. Therefore, in the process of numerical simulation research, it is necessary to comprehensively consider the calculation time and accuracy to reasonably divide the grid. Since the rotary jet tool rotates at a predetermined speed while the wellbore remains stationary during the rotary jet plug removal operation, the grid needs to be divided according to the model area, and the rotary area and the static area need to be divided, respectively. For areas with regular shapes, such as wellbore, we used structured grids to divide them, while for other areas, such as the nozzle, we used structured and unstructured grids to divide them. In this paper, the nozzle tool part was divided into 196,205 grids. At the same time, in order to improve the calculation accuracy of the model, we partially densified the grid at the nozzle, as shown in Figure 12.

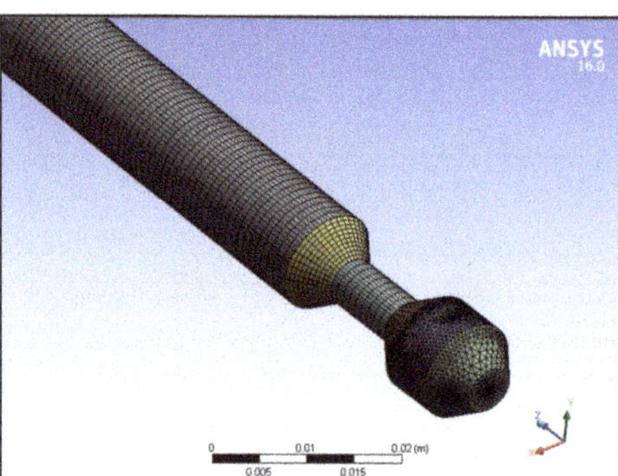

Figure 12. Meshing schemes of the established model.

3.4. Flow Field Characteristics

According to the flow field simulation results, the rotary jet could cover the whole wellbore sufficiently within a short period of time at the beginning of flushing (Figure 13). This showed that the rotary jet could fully and adequately clean the clogged and scaled locations in the wellbore with the tool in tow. By intercepting the flow field at a certain moment, it could be seen that the jet at this time has a good spot-flushing effect on the wellbore. However, due to the effect of the jet by the forward nozzle, the energy of the jet will be lost quickly in a short distance. Therefore, in practical applications, the volume flow should be increased to some extent, the spray distance should be controlled, and the full flushing of the wellbore should be achieved by dragging the tool up and down to improve the descaling rate.

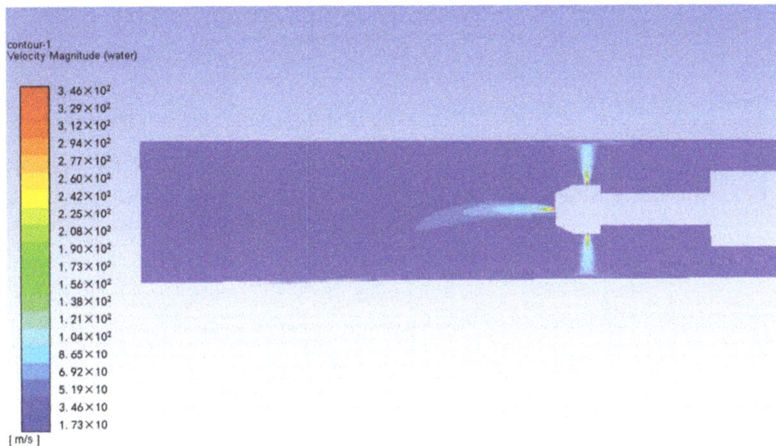

Figure 13. Schematic diagram of rotary jet's flushing effect.

In order to study the flow field distribution of the rotary jet in the wellbore, we first simulated the flow field of the rotary jet in the wellbore 1.0 s after the start of the jet, of which the results are shown in Figure 14. From the simulation results, the rotary jet has fully covered the entire wellbore within a short time of the start of the jet. Therefore, the method of dragging the rotary jet tool up and down is proposed in this paper. The method of unblocking and descaling the wellbore and near-wellbore zone through a rotary jet is practical and feasible. However, from the initial calculation example, the energy loss of the rotary jet in the wellbore is very fast. When the rotary jet reaches the wellbore, the jet velocity drops considerably, and the jet spreads gradually. It is also necessary to study the influence factors such as spray distance by changing the tool size.

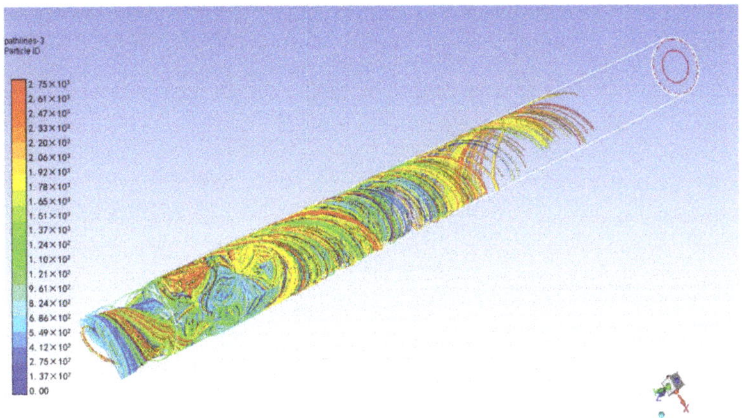

Figure 14. Schematic diagram of rotary jet's flow field.

3.5. Influence of Spray Distance

When the tool is engaged in the rotary jet unblocking operation, it needs to drive the nozzle at a certain speed and at a suitable distance to carry out jet flushing to remove the scale from the inner wall of the wellbore. From the previous numerical simulation results, it can be seen that the jet energy changes rapidly in the wellbore, which showed that the spray distance is an important factor affecting the effect of unblocking and scale removal. In order to achieve the best jet impact effect, the geometric model needs to be modified, so this section first carries out numerical simulation research around the influence of spray distance on the jet effect.

On the basis of previous studies on blockage removal and impact efficiency of the rotary jet [28,29], the spray distance is divided into 5.0 mm, 7.0 mm, 10.0 mm, 12.0 mm and 15.0 mm for research in this section. In order to study the influence of spray distance on the flow field of the rotary jet, the bottom nozzle and oblique nozzle on the rotary jet nozzle are canceled in this section to reduce the interference to the lateral rotary jet in the simulation process. The geometric model after changing the spray gun settings and spray distance is shown in Figure 15. The model area mainly includes the spray gun body, rotation area and wellbore area.

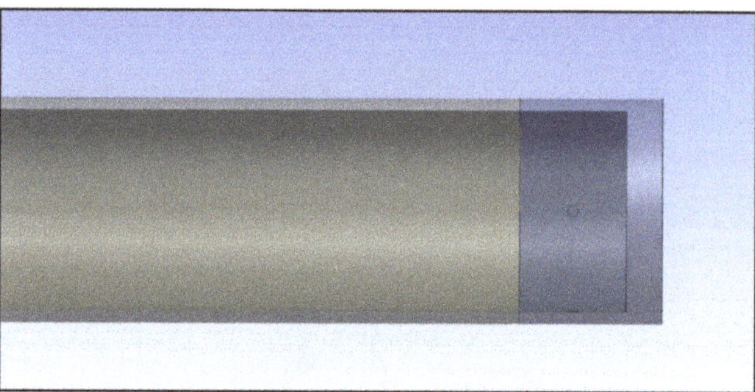

Figure 15. Schematic diagram of geometric model.

Through the simulation of the rotary jet flow field in the wellbore annulus 1.0 s after the start of the jet, it can be seen from the numerical simulation results that when the jet pitch of the rotary jet is 5.0 mm, the rotary jet can still effectively cover the wellbore area from the results of the rotary jet trajectory diagram. Therefore, we believe that when the spray distance is small, the rotary jet can still completely and fully clean the inner wall of the wellbore, making the jet less affected by interference.

During the rotary jet blockage removal operation, the distance of the rotary jet tool has an important influence on the blockage removal effect. When the distance is smaller, the energy attenuation of the high-pressure rotary jet when it contacts the inner wall of the wellbore is smaller, and the impact effect on the wellbore is stronger. At this time, the blockage removal and scale removal effect of the rotary jet is better. The influence of five groups of different jet pitches on the velocity distribution of rotary jets in the wellbore is studied by numerical simulation. The simulation results are shown in Figure 16.

From the simulation results, it can be seen from Figure 17 that when the spray distance is 5.0–10.0 mm, there is no significant difference in the flow field changes in the rotary jet in the wellbore with the increase in the spray distance. Therefore, the jet can effectively act on the inner wall of the wellbore within this range. However, when the spray distance is greater than 10.0 mm, it can be seen from the velocity distribution diagram of the rotary jet in the wellbore that the effect of the jet on the wellbore has been relatively attenuated, and the jet is relatively divergent before reaching the wellbore.

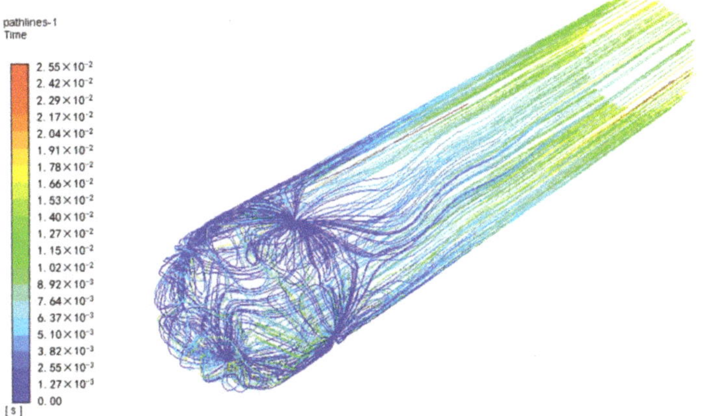

Figure 16. Trajectory map of rotary jet in the wellbore.

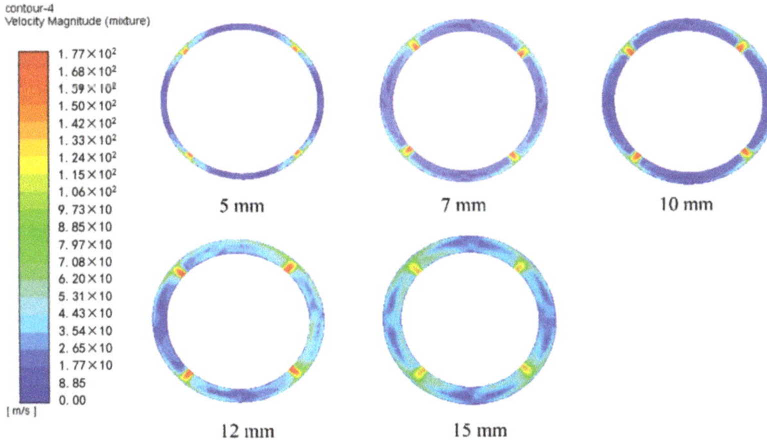

Figure 17. Velocity distribution of rotary jets at different spray distances.

Additionally, the influence of spray distances on jet pressure at the wall was studied, of which results are shown in Figure 18. It can be seen from the pressure change in the rotary jet at the well wall that the rotary jet pressure at the well wall decreases gradually with the increase in the spray distance. When the spray distance is less than 10 mm, the impact pressure of the jet on the wellbore does not drop significantly. When the spray distance is greater than 10 mm, the jet pressure at the borehole wall decreases rapidly with the increase in spray distance.

Through the analysis of the simulation results, it can be concluded that the jet energy loss will become greater with the increase in spray distance when the rotary jet reaches the inner wall of the wellbore during the plug removal operation. Therefore, during the rotary jet blockage removal operation, the jet spacing can be controlled within 5.0–10.0 mm according to the simulation results. At the same time, centralizers can be installed at the upper and lower ends of the tool string according to the actual well conditions to prevent the tool from jamming and blocking during the blockage removal operation. Reasonable jet spacing can be selected according to the actual well conditions to achieve the best blockage removal effect.

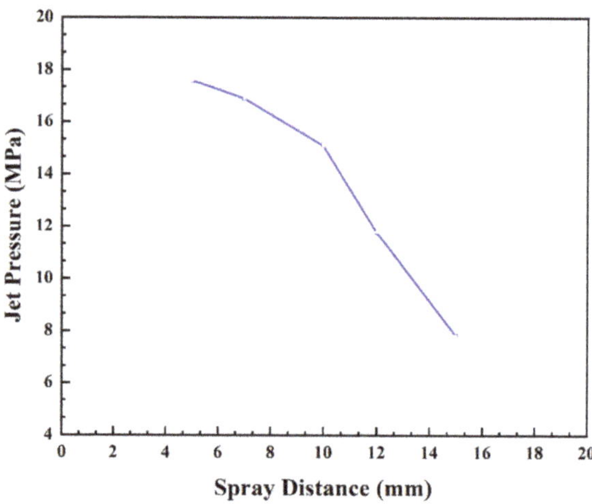

Figure 18. Variation in jet pressure at the well wall under different spray distances.

3.6. *Influence of Volume Flow*

During the rotary jet unblocking operation, the volume flow directly affects the jet pressure during the unblocking operation. Generally, the larger the construction volume flow is, the higher the nozzle pressure drop is, and the greater the impact of the rotary jet on the inner wall of the casing is, which has a significant impact on the unblocking operation effect. However, the more the volume flow increases, the faster the pump pressure will increase correspondingly and the higher the risk of unblocking operation. Figure 18 shows the velocity distribution law of the rotary jet in the wellbore under different volume flow rates.

It can be seen from Figure 19 that when the volume flow is less than 0.8 m^3/min, the rotary jet has little effect on flushing the inner wall of the wellbore, and most of the jet energy has been lost before reaching the wall. When the volume flow reaches 1.0 m^3/min, it can be seen that the central velocity of the rotating jet is relatively higher, which can be considered that the jet flushing efficiency is the highest at this time, and the energy loss of the jet when it contacts the inner wall is also smaller. When the volume flow continues to increase, even though the jet energy of the rotary jet is higher when it is greater than 1.0 m^3/min, the difference between the velocity distribution of the rotary jet in the wellbore and that of the rotary jet under the condition of 1.0 m^3/min is not obvious, indicating that the flushing efficiency of the rotary jet is not significantly improved at this time. In the process of rotary jet unblocking operation, with the increase in volume flow, the pump pressure also increases, bringing higher construction risks.

Additionally, the influence of volume flow on jet pressure at the wall was studied. The results are shown in Figure 20. It can be seen from the pressure change in the rotary jet at the well wall that the rotary jet pressure at the well wall increases gradually with the increase in the volume flow. When the volume flow is less than 1.0 m^3/min, it can be seen that with the increase in the volume flow, the jet pressure at the well wall increases rapidly. When the volume flow is greater than 1.0 m^3/min, the growth rate of jet pressure at the well wall decreases significantly. Combined with the flow field simulation results, we can suggest that the volume flow should be controlled at 1.0 m^3/min in the actual project.

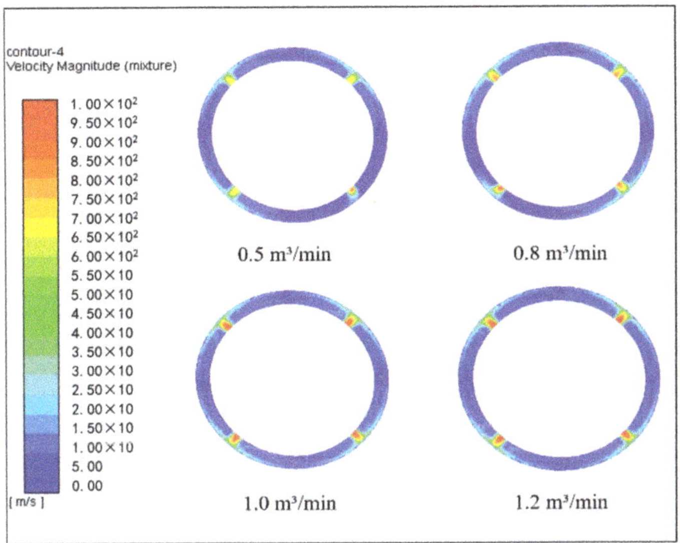

Figure 19. Velocity distribution of rotary jets at different volume flow.

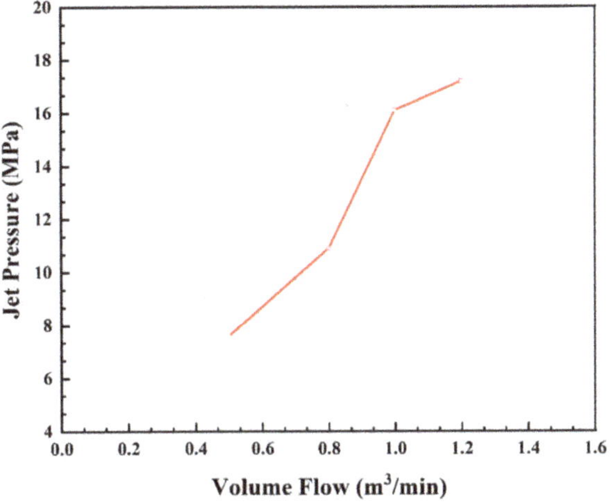

Figure 20. Variation in jet pressure at the well wall under different volume flow.

Therefore, in the actual application process, it is necessary to control the volume flow within an appropriate range to prevent high-pressure operation accidents or attenuation of volume flow from high construction pressure, which leads to the reduction in rotary jet energy, resulting in insufficient blockage removal and scale removal effect. Additionally, in order to further improve the blockage removal and scale removal effect of the rotary jet, the nozzle parameters can be designed, the appropriate nozzle combination can be used, and the operation volume flow and pressure can be appropriately increased during the blockage removal operation to achieve better blockage removal and scale removal effect.

By analyzing the change in the jet pressure at the wellbore with the spray distance and volume flow and combining it with the analysis of the flow field of the rotary jet, we can propose parameter suggestions in the actual process of the rotary jet blockage removal. Control the spray distance and volume flow in the actual operation within the appropriate range to achieve the best blockage removal effect.

4. Field Application of Rotary Jet Blockage Removal in Renre 4 Well

The aim of this section is to verify whether the high-pressure rotary water jet can effectively remove the blockage and scale of the low-efficiency wells in the sandstone thermal reservoir of the Guantao Formation and improve the permeability near the well and increase the reinjection volume. In this chapter, a low-efficiency reinjection well named Renre 4 Well, located in the Bohai Vocational College area, was selected for field application. The corresponding rotary jet unblocking construction scheme was designed, and the on-site unblocking operation was carried out. At the end of the operation, the effect of blockage removal and irrigation increase in the field application was evaluated.

4.1. Design of the Rotating Jet Unblocking Tool

In this section, based on the principle of plug removal and permeability enhancement of rotary jet for the Guantao Formation sandstone formation, combined with the results of numerical simulation and jet experiment, we designed the rotary jet plug removal tool for the problem of plug removal and permeability enhancement of low-efficiency wells in sandstone thermal reservoir, and processed the rotary jet plug removal tool for Guantao Formation sandstone thermal reservoir, providing a tool basis for on-site plug removal of the rotary jet.

The specific working principle of plug removal is as follows: before the plug removal operation, the rotary jet plug removal tool is lowered to the predetermined cleaning layer through the oil tube, and the clean water enters the jet plug removal tool through the oil tube, filter and bypass valve through the pump truck to generate multiple high-pressure water jets. At the same time, the plug removal tool nozzle is driven to rotate in the wellbore through the lifting and lowering of the string and the friction between the elastically retractable inclined roller and the inner wall of the casing. The high-pressure water jet generated by the blockage removal tool can clean the whole wellbore through direct flushing, high-frequency oscillating hydraulic wave and cavitation noise and effectively physically remove the blockage and scaling substances in the wellbore and gun hole.

The rotary jet tool has a wide range of applications. The size of the tool body and the nozzle layout of the spray gun body can be adjusted according to different working conditions and well structure. The rotary jet plug removal tool is applicable to geothermal wells or oil and gas wells where the reinjection volume is reduced due to pollution or blockage near the well zone. Geothermal wells or oil and gas wells with high formation acid sensitivity and unsuitable for acidizing and other stimulation measures. Wells have a thin reservoir and small interval, which is not easy when taking other reconstruction measures. The rotary jet tool can be used for pre-treatment before acidizing, fracturing, sand control and other measures to improve the downhole environment.

The rotary jet unblocking tool is composed of a high-pressure rotary joint, spray gun body, nozzle, rotary device, centralizer, bearing and damper. When working, the upper part of the tool is connected to the bypass valve, filter, safety connector, etc., to form a rotating jet unblocking tool string; it shall be sent to the predetermined layer for blockage removal with tubing, the ground cement truck starts the pump to pressurize, and the clean water or the working fluid added with an anti-swelling agent, wax inhibitor and clay stabilizer passes through the oil tube, filter, bypass valve (one-way valve) and then enters the incident flow unblocking tool to generate multiple radial high-pressure water jets. At the same time, the ground operation vehicle slowly lifts the tubing string to make the tool rotate while moving up and down in the cleaning section.

As the sandstone thermal reservoir of the Guantao Formation is poorly consolidated and the formation is relatively loose, Renre 4 well is completed by perforation, and the inner diameter of the casing is 226.0 mm. Therefore, to ensure the effect of blockage removal and permeability enhancement, in this chapter, in combination with the research results of numerical simulation and the experimental results of high-pressure jet blockage removal, we processed the maximum outside diameter of the rotary jet tool to 215.0 mm, and the total length of the tool was 1947.0 mm. By installing 218.0 mm centralizers at the upper and lower ends of the tool, the rotary jet unblocking tool can ensure that the spray distance of the rotary jet nozzle can be kept within the range of 5.0–10.0 mm during the unblocking operation so as to ensure the effect of unblocking and permeability enhancement.

In this chapter, the tool spray gun body is arranged with 12 radial nozzles, which are divided into two layers. Six nozzles are evenly distributed on the spray gun body in each layer. Different nozzle combinations can be changed according to the operation requirements. The nozzle can be replaced with a combination of 2.6 mm, 3.0 mm and 4.0 mm nozzles to improve the flushing efficiency during the unblocking operation. The total length of the plug removal tool string is 2.94 m, of which the tool body is about 1.2 m long, the maximum rigid outer diameter is 215 mm, and the maximum working pressure can reach 40 MPa. The rotary jet blockage removal tool meets the requirements of the rotary jet blockage removal operation for the low-efficiency wells of sandstone geothermal reservoirs and provides the tool conditions for the field operation of rotary jet blockage removal in this chapter.

Figure 21 shows the rotary jet plug removal tool string processed in the field plug removal research in this chapter. After installation, the rotary jet plug removal tool string consists of a centralizer, rotary device, spray gun body, rotary joint, short filter joint and safety joint. The rotary joint and short filter joint are connected by a ball socket. Before the plug removal, put the steel ball into the ball socket from the wellhead, and the safety joint is connected with a 2–7/8″ oil tube, and then put the tool into the wellhead.

Figure 21. Installation diagram of rotary jet unblock tool.

4.2. Construction Process and Parameters

In this section, we carried out the rotary jet unblocking operation of Renre 4 Well in the western region of Renqiu (Figure 22) and carried out the continuous rotary jet unblocking operation for the predetermined unblocking zone of Well Renre 4. The construction process parameters and application effects are as follows.

In the design of the rotary jet unblocking process for Renre 4 Well, the pump pressure is maintained below 25 MPa during construction, the construction volume flow is maintained at 1.2 m^3/min, the total length of the unblocking operation interval of Renre 4 Well is 334 m. During the construction process, a single oil tube is lifted and lowered three times. There are 22 unblocking intervals in the construction process, and the total duration of the unblocking operation is 5.2 h. The specific unblocking data are shown in Table 2.

Figure 22. The process of Renre 4 well unblocking operation.

Table 2. Unblocking operation parameters of Renre 4 well.

Construction Layer m	Interval Number /	Pressure MPa	Volume Flow m^3/min
1784–1931	14	18.4–20.8	1.18–1.21
1529–1716	8	18.8–21.7	1.20–1.26

4.3. Results and Analysis of Field Application

In the process of rotary jet unblocking operation of Renre 4 Well, there is less fluid returning at the wellhead at the initial stage of unblocking operation. With the unblocking operation, the fluid return gradually occurs at the wellhead, which is turbid but has less sand content. At the later stage of the plug removal operation, there is a certain amount of sand in the wellhead return fluid. During the well flushing operation, after the completion of the plug removal operation, much sand is discharged from the circulating fluid return. The fluid return and sand production are shown in the figure below. Most of the fluid return and sand production are fracturing sand used in fracturing construction; this is consistent with the blockage analysis results of Renre 4 Well in the previous study. After the blockage removal operation, the bottom hole condition and reinjection condition of Renre 4 Well were also improved significantly, as shown in Figure 23 below.

Figure 23. Returning fluid and sand output during the flushing process.

By the end of 2021, the reinjection volume of Renre 4 Well before and after plug removal was 8–9 m^3/h. The amount of reinjection has decreased to a lower level. After the rotary jet unblocking operation, the reinjection volume of Renre 4 Well can reach 14 m^3/h. When the pressure is 0.6 MPa, the reinjection volume can reach 15 m^3/h. The total amount of reinjection increased by 55.6–87.5%. At the end of construction, after the well flushing tool string is taken out, much asphaltene is attached to the tool surface, but the spray gun body and nozzle are kept intact, which proves that the rotary jet blockage removal tool meets the construction requirements of high-pressure rotary jet blockage removal. After the rotary jet blockage removal operation, the reinjection volume of Renre 4 Well was also effectively improved, which verifies the effectiveness of the rotary jet blockage removal and injection increase for the sandstone geothermal reservoir of Guantao Formation and the feasibility of the method.

5. Conclusions

The key findings of the study are as follows:

(1) Experiments on the transmission efficiency of jet pressure waves were designed to study the effects of high-pressure jets on screen pipe blockage removal and scale removal, as well as the erosion and fragmentation of sandstone samples. The experimental results of jet pressure wave transmission efficiency show that after continuous flushing by a high-pressure jet, the blockage and rusting state of the inner wall of the screen pipe has been significantly improved. Additionally, after passing through the screen pipe, the high-pressure jet has a certain erosion and crushing effect on the yellow sandstone on the outer wall of the wellbore. The feasibility of the rotating jet unblocking method is verified.

(2) The sandstone in Guantao Formation has poor consolidation and is loose and easily broken. Based on the experiment results, it is suggested that the conditions such as flow rate and spray distance should be controlled according to the actual situation of the formation in the process of rotary jet blockage removal to prevent damage and destruction of the formation near the well. At the same time, gas lift and other well flushing operations are required to fully clean the wellbore after the completion of jet flow.

(3) Based on the blockage removal principle of rotary jet, this paper conducted a numerical simulation study on the flow field of the rotary jet in the wellbore, which verifies the feasibility of blockage removal and enhanced injection of rotary jet for low-efficiency wells in sandstone geothermal reservoirs. The influence of spray distance and volume flow on the rotary jet was studied. Based on the numerical simulation results, suggestions on spray distance and volume flow parameters in the process of rotary jet blockage removal are proposed.

(4) The rotary jet unblocking tool and corresponding construction technology were designed, and the rotary jet unblocking construction scheme was prepared for Renre 4 Well. After the unblocking operation, the blockage of the target well was effectively improved, and the reinjection volume flow was increased. After analyzing the sediment by sampling the flowback liquid, the results are consistent with the previous analysis of blockage causes. After the completion of the blockage removal operation, the spray gun body and nozzle of the rotary jet plug removal worker are in good condition, which also proves that the set of rotary jet blockage removal tools can meet the blockage removal requirements of the low-efficiency wells in the geothermal reservoir of the sandstone of the Guantao Formation, providing some guidance on the application of the rotary jet technology in the blockage removal and permeability enhancement of the sandstone geothermal reservoir.

(5) In this paper, the application and promotion of rotary jet unblocking technology in sandstone geothermal reservoirs are provided. However, the rotary jet can be applied to other types of geothermal reservoirs, such as carbonate rock or other geothermal reservoirs, in future work. For example, the rotary jet blockage removal technology can adapt to the unblocking requirements of different types of geothermal reservoirs by changing the structure and combination of tool nozzles and combining different types of working fluids.

Author Contributions: Conceptualization, C.Y.; methodology, C.Y.; software, T.T.; formal analysis, C.Y., C.H. and T.T.; investigation, Y.Z.; writing—original draft preparation, C.Y.; writing—review and editing, C.Y., C.H. and H.H.; supervision, Y.Z. All authors have read and agreed to the published version of the manuscript.

Funding: This research was funded by the National Key Research and Development Program of China, grant number 2019YFB1504202.

Data Availability Statement: Not applicable.

Conflicts of Interest: The authors declare no conflict of interest.

References

1. Lund, J.W. Direct heat utilization of geothermal resources. *Renew. Energy* **1997**, *10*, 403–408. [CrossRef]
2. Fridleifsson, I.B. Geothermal energy for the benefit of the people. *Renew. Sustain. Energy Rev.* **2001**, *5*, 299–312. [CrossRef]
3. Hou, J.; Cao, M.; Liu, P. Development and utilization of geothermal energy in China: Current practices and future strategies. *Renew. Energy* **2018**, *125*, 401–412. [CrossRef]
4. Javadi, H.; Ajarostaghi, S.S.M.; Rosen, M.A.; Pourfallah, M. Performance of ground heat exchangers: A comprehensive review of recent advances. *Energy* **2019**, *178*, 207–233. [CrossRef]
5. Wang, G.; Song, X.; Shi, Y.; Zheng, R.; Li, J.; Li, Z. Production performance of a novel open loop geothermal system in a horizontal well. *Energy Convers. Manag.* **2020**, *206*, 112478. [CrossRef]
6. Junhong, A. Present situation and development prospect of geothermal tailwater reinjection technology in plateau. *China Plant Eng.* **2017**, *24*, 137–138.
7. Kaya, E.; Zarrouk, S.J.; O'Sullivan, M.J. Reinjection in geothermal fields: A review of worldwide experience. *Renew. Sustain. Energy Rev.* **2011**, *15*, 47–68. [CrossRef]
8. Song, X.; Zheng, R.; Li, G.; Shi, Y.; Wang, G.; Li, J. Heat extraction performance of a downhole coaxial heat exchanger geothermal system by considering fluid flow in the reservoir. *Geothermics* **2018**, *76*, 190–200. [CrossRef]
9. Seibt, P.; Kellner, T. Practical experience in the reinjection of cooled thermal waters back into sandstone reservoirs. *Geothermics* **2003**, *32*, 733–741. [CrossRef]
10. Ungemach, P. Reinjection of cooled geothermal brines into sandstone reservoirs. *Geothermics* **2003**, *32*, 743–761. [CrossRef]
11. Song, W.; Liu, X.; Zheng, T.; Yang, J. A review of recharge and clogging in sandstone aquifer. *Geothermics* **2020**, *87*, 101857. [CrossRef]
12. Siriwardene, N.R.; Deletic, A.; Fletcher, T. Clogging of stormwater gravel infiltration systems and filters: Insights from a laboratory study. *Water Res.* **2007**, *41*, 1433–1440. [CrossRef]
13. Ma, Z.; Hou, C.; Xi, L.; Yun, H.; Sun, C. Reinjection clogging mechanism of used geothermal water in a super-deep-porous reservoir. *Hydrogeol. Eng. Geol.* **2013**, *40*, 133–139.
14. Houben, G. Iron oxide incrustations in wells. Part 1: Genesis, mineralogy and geochemistry. *Appl. Geochem.* **2003**, *18*, 927–939. [CrossRef]
15. Liu, X.L.; Zhu, J.L. A study of clogging in geothermal reinjection wells in the Neogene sandstone aquifer. *Hydrogeol. Eng. Geol.* **2009**, *36*, 138–141.
16. Liu, G.; Wu, S.; Fan, Z.; Zhou, Z.; Xie, C.; Wu, J.; Liu, Y. Analytical Derivation on Recharge and Periodic Backwashing Process and the Variation of Recharge Pressure. *J. Jilin Univ. (Earth Sci. Ed.)* **2016**, *46*, 1799–1807.
17. Xia, L.; Zheng, X.; Shao, H.; Xin, J.; Sun, Z.; Wang, L. Effects of bacterial cells and two types of extracellular polymers on bioclogging of sand columns. *J. Hydrol.* **2016**, *535*, 293–300. [CrossRef]
18. Chen, J.; Yang, R.; Huang, Z.; Li, G.; Qin, X.; Li, J.; Wu, X. Detached eddy simulation on the structure of swirling jet flow field. *Pet. Explor. Dev.* **2022**, *49*, 806–817. [CrossRef]
19. Zhang, Y.; Wu, X.; Li, G.; Hu, X.; Hui, C.; Tan, Y.; Huang, H. Study on erosion performance of swirling cavitating jet for natural gas hydrate. *J. Cent. South Univ. (Sci. Technol.)* **2022**, *53*, 909–923.
20. Felli, M.; Falchi, M.; Pereira, J.A. Distance effect on the behavior of an impinging swirling jet by PIV and flow visualizations. *Exp. Fluids* **2010**, *48*, 197–209. [CrossRef]
21. Li, G.; Shen, Z.; Peng, Y. A theoretical study of hydraulically rotating jet nozzle. *Acta Pet. Sin.* **1995**, *16*, 148–153.
22. Bu, Y.; Wang, R.; Zhou, W. Study on flow principles of rotating jet. *Oil Drill. Prod. Technol.* **1997**, *19*, 7–10+105.
23. Li, J.B.; Li, G.S.; Huang, Z.W.; Song, X.Z.; Li, K. Flow Field Study on a New Kind Swirling Multi-orifices Nozzle. *Fluid Mach.* **2015**, *43*, 32–36+41.
24. Percin, M.; Vanierschot, M.; Van Oudheusden, B. Analysis of the pressure fields in a swirling annular jet flow. *Exp. Fluids* **2017**, *58*, 166. [CrossRef]
25. Li, J.; Li, G.; Huang, Z.; Song, X.; He, Z. Effect of confining pressure on the axial impact pressure of hydraulic jetting. *J. Exp. Fluid Mech.* **2017**, *31*, 67–72.
26. Liang, X.; Dezu, R.; Yonghao, M.; Jin, L.; Jianmin, G. Pressure loss and heat transfer characteristics experiment of swirling impinging jet with different shape nozzles. *J. Aerosp. Power* **2018**, *33*, 2678–2686.

27. Hanjie, C.; Zhenping, S.; Yuan, L.; Aibaibu, A.; Yuezhong, W.; Xiaowei, W. Numerical Simulation of Water Jet Perforation on Casing Surge Pressure. *Oil Field Equip.* **2020**, *49*, 45–48.
28. Ivanic, T.; Foucault, E.; Pecheux, J.; Gilard, V. Instabilities in coaxial rotating jets. *J. Therm. Sci.* **2000**, *9*, 322–326. [CrossRef]
29. Love, T.; McCarty, R.; Surjaatmadja, J.; Chambers, R.; Grundmann, S. Selectively placing many fractures in openhole horizontal wells improves production. *SPE Prod. Facil.* **2001**, *16*, 219–224. [CrossRef]

MDPI
St. Alban-Anlage 66
4052 Basel
Switzerland
Tel. +41 61 683 77 34
Fax +41 61 302 89 18
www.mdpi.com

Energies Editorial Office
E-mail: energies@mdpi.com
www.mdpi.com/journal/energies

www.ingramcontent.com/pod-product-compliance
Lightning Source LLC
LaVergne TN
LVHW071000170726
843515LV00006B/1032

* 9 7 8 3 0 3 6 5 6 4 0 2 9 *